Documentation Systems

Clear and Simple

James L. Vesper

Interpharm Press, Inc.
Buffalo Grove, Illinois

Invitation to Authors

Interpharm Press publishes book focused upon applied technology and regulatory affairs impacting Healthcare Manufacturers worldwide. If you are considering writing or contributing to a book applicable to the pharmaceutical, biotechnology, medical device, diagnostic, cosmetic, or veterinary medicine manufacturing industries, please contact our Director of Publications.

Social Responsibility Programs

Reforestation

Interpharm Press is concerned about the impact of the worldwide loss of tress upon both the environment and the availability of new drug sources. Therefore, Interpharm supports global reforestation and commits to replant trees sufficient to replace those used to meet the paper needs to print its book.

Pharmakos-2000

Through its Pharmakos-2000 program, Interpharm Press fosters the teaching of pharmaceutical technology. Under this program, complimentary copies of selected Interpharm titles are regulatory sent to every College and School of Pharmacy worldwide. It is hoped that these books will be useful references to faculty and students in advancing the practice of pharmaceutical technology.

10 9 8 7 6 5 4 3 2 1
ISBN: 1-57491-050-7

Interpharm Press, Inc.
1358 Busch Parkway
Buffalo Grove, IL 60089, USA
Ph: 1+847+459-8480 Fx: 1+847+459-6644

Contents

PREFACE

"If it isn't written down, it's just a rumor."

That quote, repeatedly given by FDA and quality assurance professionals, is the underlying reason for this book: to be in a state of control, instructions and requirements must be formally written down. Records that can help someone reconstruct an event or decision need to be properly prepared.

Documentation is the cornerstone of a quality system, whether the system is based on Good Manufacturing Practices or ISO-9000. Not only do you need to have documents like specifications, protocols, and procedures, these functional documents must be developed, distributed, and maintained using a formalized process that is carefully controlled.

If you've been lucky to be audited by a governmental revenue agency (like the United States' Internal Revenue Service or Revenue Canada), you probably have experienced their expectations for records. Well organized, detailed receipts are a taxpayer's best friend during an audit. If it isn't written down, you have no way to show it really happened. The same holds true for records in developing, manufacturing, and testing medical devices, drugs, and biologicals. These document must clearly explain what was done and justify the release of the product into the marketplace.

This was a difficult book to write. Or, more precisely, it was difficult figuring out what *not* to write in this book. The subject of documentation and recordkeeping, as it applies to the pharmaceutical and medical device industries could easily consume several volumes.

This book is based on a workshop that I have given, to firms and groups throughout the United States, Canada, Asia, and Europe. In many of these courses, I have had the opportunity to collaborate with two friends and colleagues, Art Mlodozeniec and Farrow Beacham. Each time we have given the workshop, we've covered a core set of content, but then, based on the participants and their desires, we go off in a variety of directions, covering things from LIMS systems to the role of management in documentation programs.

This book is written for:

- People new to the area of documentation;
- People new to the pharmaceutical, medical device, and biologics industries;
- People who are not documentation specialists but need to audit such areas; and
- Practitioners in the field who require a reference.

The selection of content was made with an emphasis on the systems aspect of documentation and recordkeeping that can be applied in the widest number of situations.

As you read, you may discover two biases of mine. First, that documentation and recordkeeping are not and should not be viewed as a pain in the neck, Rather, they are things that should help the organization accomplish its goals. The second bias is that functional documents, particularly procedures and methods, need to be created with thought given at the outset to training.

One caution with this book—it is the same caution we give to the workshop participants. You will not find any magic answers here. You won't find ways to cut your document library by one-half, or reduce the number of documentation coordinators at your firm by 40 percent. You may, in fact, discover you need *more procedures* or that *more people* have to be involved in the process. Since documentation systems are one of the most complex components of your firm's quality system, this must be expected. Your goal, however, must be constantly kept in mind: to help capture the knowledge of your firm and replace rumors with facts.

Jim Vesper

July 1997

Acknowledgments

In addition to the friends, family, and colleagues who once again showed their understanding while I worked on this, another book project, there are several other people who need special recognition. Arthur Mlodozeniec and Farrow Beacham were co-instructors with me as we presented workshops on Documentation Systems. Some of their content is included in this volume, as well as ideas that were developed and refined in our many conversations. Thanks also to Sara Krusenstjerna for reviewing some of the content. Amy Davis, of Interpharm Press, is another important person in getting this book together. Her patience, understanding, creative solutions, and encouragement helped to make this happen.

Design and Layout:	Chris Clark
Cover Design:	Michelle Renaud, Level 3, NTID Applied Art Computer Graphics Student

Dedication

To my parents

CHAPTER 1

Why Document?

Goals

- Define *document*.
- Identify and discuss the business, quality, technical, and regulatory rationale for documents.

Why document? Why write something down?

You might have played the game called "pass the message" or "telephone" when you were in grade school. In a workshop on Documentation Systems we play it as a way for people to examine the reasons for documentation.

Start off with any message: *Lunch on Tuesday will be served in the Golden Glade dining room at 12:30. It will include leek soup and turkey sandwiches on rye bread.*

Have a person whisper it once to the next person, going around the room with the last person writing down what he or she hears. The more people, the more varied their native languages, the more interesting the results:

"Lunch will be leaking turkey."

"Lunch is at 11:30 today."

"Gold soup will be given in paper bags."

This simple activity demonstrates some of the problems with oral communication.

- Spoken information is prone to be modified: there is no reference standard that can be referred to at a later time.
- The changes in the message are difficult to reconstruct at a later point in time: who changed the location or why someone left out a particular phrase.
- Oral communication is subject to interference: noise can block a message or garble particular details. The more complex the message, the higher the potential for losing specific elements.
- Oral communication is extremely transitory: it is preserved only in the minds of the transmitter and receiver(s). It will be saved as a memory, but that can be affected by time or colored by other events.

For these reasons, oral communication is not a preferred way of providing instructions or recording what happened. If information needs to be given to others, if it must be consistent and definitive, if it needs to be used at other times and places, it must be documented.

Before going on with other reasons for documenting, what exactly is a *document*? Traditionally, a document has been defined as a written or printed paper bearing the original, official, or legal form of something. In our multimedia age, however, a document can be more than just symbols on paper. A voice message or videotape can be a document. A document could exist as digital code on the hard drive of a computer.

For our purposes, a document is *an organized form of information that can serve as legal evidence or proof*. While we will be emphasizing printed words or symbols, most of our discussion is equally valid for information in other formats. Indeed, as new technologies appear that make it more efficient to create, transmit, and store moving images, issues like review, approval, change control, etc. will become more and more important.

With this working knowledge of why we document and what a document is, what are some other reasons for documenting?

Business Reasons for Documenting

Documents are a valuable resource

More than ever before, information is a strategic resource. For example, a development report describes things learned in the development of a new drug product or medical device. It identifies critical production parameters that will be validated, describes failures and what caused them. In total, this report summarizes years of research in a well-organized way. Firms that excel at technology transfer and getting products quickly onto the market have learned that development reports are a key tool, a treasure of crucial information.

A pharmaceutical firm that went through a downsizing in the early 1990s discovered the costs of *not* having information written down. Much of the "art" in making a particular product resided only in the heads of an engineer. The engineer, who took early retirement, was hired back as a consultant with the task of writing detailed procedures describing the process and methods. (Fortunately for him, he was making considerably more money in his new consulting role than he was as a regular employee.)

Documents are used for communication

Symbols on paper can be much more precise and less ambiguous than just an oral description. Some symbols work better in some situations than in others. For instance, it may be much simpler to assemble a medical device using a drawing than following a narrative description. Decision trees, a form of flow diagram, can help users make more consistent choices than a long paragraph that describes when certain options are to be used.

Documents promote consistency and standardization. This is especially important when a task is done repeatedly and/or by different people in various locations. Documents not only can describe the standards that are expected, they help people accomplish the goal by providing a common path that all will follow.

Documents help with decision making

Reports can present facts that can be reacted to. Assuming the information is correct and objectively presented, decision-makers can use documents as a basis for their decisions. They can repeatedly refer to the information as they decide on a particular course of action.

Documents contain history

Documents can provide a historical record of how something was done or the rationale behind a decision. For centuries, researchers have been struggling to discover the secret that makes the sound from a Stradivarius violin so special. Could it be the way the wood was dried or the source of the shellac? Nobody really knows for sure, partly because none of this was documented. A medical device firm observed that in a recent ten-year period, seven different teams tried to create the same component and failed. No team knew about the other; the teams spent millions of dollars reaching the same point of failure.

Scientific Reasons for Documenting

Documents preserve data and observations

Examining old scientific notebooks can be fascinating, as you are coming into direct contact with the day-to-day lives of great minds. For example, some of the notebooks of Leonardo de Vinci are works of art themselves. Ideas, failures, detours, and successes are recorded here. The more clearly they are written, either with formulae, words, or drawings, the easier it is for other knowledgeable personnel to understand the original experiments and results along with their significance.

Documents substantiate patent claims

Substantiating the date of invention is one of the most important uses of a laboratory notebook, at least from an economic point of view. Patents are awarded to the person or organization who can support the earliest date of invention. Fame (and fortune) depend not only on proof of conception of the idea, reduction to practice, and diligence, the notebook and recordkeeping system are needed to refute legal challenges to their integrity.

Documents provide a forum for analysis and interpretation

Simply writing something down in a laboratory notebook helps the writer's brain organize and process the information. Graphical forms of the data can show relationships more easily that might not be noticed if someone were to just observe readings on a display device.

Another reason to document . . . it may help you lose weight.

"We all get heavier as we get older because there's a lot more information in our heads."

Los Angeles Laker's center Vlade Divac, on why he showed up for training camp 15 pounds heavier than he was the previous season (*Newsweek*, Nov 23, 1992).

Quality Reasons for Documenting

All quality management programs, whether an in-house version based on the principles of quality gurus Deming or Juran or the ISO-9000 quality system, require there be documentation and records.

While ISO-9000 certifiers seem not as concerned with the technical details of what is in various pieces of documentation, they do expect that the system be structured properly and well controlled.

Following the Deming Cycle of "plan ➡ do ➡ check ➡ act" requires that you somehow keep track of the results and produce a written plan that is acted upon and later compared to the actual results.

Regulatory Reasons for Documenting

In developing, manufacturing, and testing drugs and medical devices, various regulations require an organization to keep records. You don't have a choice: *to stay in business, you need to do these things*. In addition to the documents and records specifically required in the regulations, documented evidence that you are following the regulations will be required by the inspectors. A challenge to you as a documentation professional is to get as much value from a business, scientific, and quality perspective as possible. (The next chapter examines some of the specific requirements in various regulations.)

From a GMP perspective, agency personnel, when doing an on-site inspection, will want written, documented answers to these four basic questions.

1. ***What are you doing?***

   Examples of documents expected: standard operating procedures, protocols, master formulae, work instructions, methods, specifications, master records, and assembly drawings.

2. ***Why are you doing it?***

   Examples of documents expected: validation reports, development reports, rationales, investigation reports, policies.

3. ***Are you doing what you have said?***

   Examples of documents expected: batch manufacturing records, batch history records, cleaning records, equipment use logs, preventative maintenance logs, in-process data, statistical process control, release data.

4. ***Are things under control?***

   Examples of documents expected: product quality reviews, internal audit reports (a document usually not read by FDA inspectors), deviation logs, compliance investigations.

One more reason to document . . . you want to be a saint.

> *The Vatican has begun procedures for naming Pope Paul VI a saint . . . . [The] cause 'is moving very swiftly,' says Jesuit Father Peter Gumpel. One reason: archives of Paul VI's papers in his hometown, Brescia, make it easier than usual to document a possibly saintly life.*
> (*Newsweek*, Jan 4, 1993)

Conclusion

As you begin to think about the documentation system at your firm, the following questions can help you focus on what its strengths and weakness are. With that information, you can begin to make your program stronger, more effective, and an efficient part of your firm's quality system.

- Are your documents helping you achieve your business goals?
- Are your documents useful from a technical and scientific point of view?
- Are your documents helping you continuously improve the products, services, and information your organization produces?
- Will your documents satisfy the expectations of the regulatory officials who review your documents?

Chapter 2

Regulatory Requirements for Documentation Systems

Goals

- Identify GMP regulatory requirements that apply to documentation systems.
- Discuss recent FDA citations in Warning Letters and 483s related to documentation systems and recordkeeping.
- Identify and discuss expectations that some FDA inspectors have that go beyond the written regulations.
- Discuss the FDA's requirements for electronic records and signatures.

A simple summary of good manufacturing practice that extends across all products and beyond regulatory boundaries are the following six points:

1. You have defined what you want to do;
2. You know that what you will do is really going to work;
3. You do what you have said you are going to do;

4. You record what you did;
5. You take the appropriate actions if anything seems wrong; and
6. The products you produce meet the highest standards of safety, identity, strength, purity, and quality.

The primary method of communicating information about what you are going to do and also showing what actually took place involves documents and records. Governments that have established regulations that apply to drugs and medical devices have expectations related to documentation and recordkeeping. This chapter examines requirements from several different sets of GMP requirements.

Regulatory Requirements

There are very few *detailed* requirements that describe in writing what the FDA and other agencies expect to see related to generating documents for pharmaceutical products. Agencies have general requirements found in the regulations as well as specific, unwritten expectations. Based on the written regulations, the U.S. Food and Drug Administration (FDA) and the Canadian Health Protection Branch (HPB) seem to be more interested in listing what documents (as a minimum) are required than in the process of producing them. On the other hand, the World Health Organization's GMP document and European Community describe what should be considered in a documentation system. These EC and WHO requirements are what other agencies, such as the FDA, expect to see implemented in the firms they inspect.

For medical devices, the FDA, in its new Quality System Regulation, provides some detail of what it expects in a documentation system, with an emphasis on who is involved with changes to documents.

GMP references specifically connected to a firm's documentation system are shown below. Tables 2.1–2.5 present the requirements from various regulatory groups. Detailed listings of the particular documents are not included; guidelines for this are found in Chapter 13. In addition, a more thorough examination of document archiving and retention is covered in Chapter 11.

Table 2.1. U.S. FDA cGMPs for Finished Drug Products, (21 CFR 211)

Citation	Requirements
§211.22	(c) The quality control unit shall have the responsibility for approving or rejecting all procedures or specifications impacting on the identity, strength, quality, and purity of the drug product.
§211.94	(d) Standards or specifications, methods of testing, . . . and [other methods] shall be written and followed. . . .
§211.100	(a) There shall be written procedures . . . . Such procedures shall include all requirements in this subpart. These written procedures, including any changes, shall be drafted, reviewed . . . by the appropriate organizational units . . . .

(Continued on next page.)

(Continued from previous page.)

§211.160	(a) The establishment of any specifications, standards, sampling plans, test procedures, or other laboratory control mechanisms . . . shall be drafted by the appropriate organizational unit and reviewed and approved by the quality control unit.
§211.186	(a) . . . master production and control records for each drug product, including each batch size thereof, shall be prepared, dated, and signed (full signature, handwritten) by one person and independently checked, dated, and signed by a second person. The preparation of master production and control records shall be described in written procedures and such written procedures shall be followed.
§211.188	Batch production and control records shall be prepared for each batch of drug product produced and shall include. . . .

Table 2.2. U.S. FDA Quality System Regulations for Medical Devices (21 CFR 820) (Published Oct 7, '96)

Citation	Requirements
§820.40	Each manufacturer shall establish and maintain procedures to control all documents required by this part. The procedures shall provide for the following:

(Continued on next page.)

(Continued from previous page.)

(a) Documentation approval and distribution. Each manufacturer shall designate an individual(s) to review for adequacy and approve prior to issuance all documents established to meet the requirements of this part. The approval, including the date and signature of the individual(s) approving the document, shall be documented. Documents established to meet the requirements of this part shall be available at all locations for which they are designated, used, or otherwise necessary, and all obsolete documents shall be promptly removed from all points of use or otherwise prevented from unintended use.

(b) *Document changes.* Changes to documents shall be reviewed and approved by an individual(s) in the same function or organization that performed the original review and approval, unless specifically designated otherwise. Approved changes shall be communicated to the appropriate personnel in a timely manner. Each manufacturer shall maintain records of changes to documents. Change records shall include a description of the change, identification of the affected documents, the signature of the approving individual(s), the approval date, and when the change becomes effective.

§820.70 (b) *Production and process changes.* Each manufacturer shall establish and maintain procedures for changes to a specification, method, process, or procedures. Such changes shall be verified or where appropriate validated . . . .

Table 2.3. Canadian HPB GMP Guidelines for Finished Drug Products (Fourth Edition)

Citation	Requirements
C.02.018	The specifications . . . shall (a) be in writing; (b) be approved by the person in charge of the quality control department; and (c) comply with the Act and these Regulations.
C.02.024 Interpretation–General	Authorized written SOPs for all sections . . . retained for reference and inspection. These SOPs are regularly reviewed and kept up to date by authorized persons. The reasons for revision are documented. A system is in place to ensure that only current SOPs are in use. All relevant GMP documents (such as associated records of actions taken or conclusions reached) and SOPs are approved, signed, and dated by authorized persons. Documents are not altered without authorization. Any alteration made to a document is signed and dated; the alteration permits the reading of the original information. Where appropriate, the reason for the change is recorded.

Table 2.4. World Health Organization (WHO) Good Manufacturing Practices for Pharmaceutical Products (1992)

Citation	Requirements
14	.2 Documents should be designed, prepared, reviewed, and distributed with care. They should comply with the relevant parts of the manufacturing and marketing authorizations.

(Continued on next page.)

(Continued from previous page.)

.3 Documents should be approved, signed, and dated by appropriate authorized persons. No document should be changed without authorization.

.4 Documents should have unambiguous contents: the title, nature, and purpose should be clearly stated. They should be laid out in an orderly fashion and be easy to check. Reproduced documents should be clear and legible. The reproduction of working documents from master documents must not allow any error to be introduced through the reproduction process.

.5 Documents should be regularly reviewed and kept up to date. When a document has been revised, a system should exist to prevent inadvertent use of the superseded version.

.6 Where documents require the entry of data, these entries should be clear, legible, and indelible. Sufficient space should be provided for such entries.

.7 Any alteration made to a document should be signed and dated; the alteration should permit the reading of the original information. Where appropriate, the reason for the alteration should be recorded.

.8 Records should be made or completed when any action is taken and in such a way that all significant activities concerning the manufacture of pharmaceutical products are traceable. Records and associated standard operating procedures should be retained for at least one year after the expiry date of the finished product.

(Continued on next page.)

(Continued from previous page.)

	.9 Data may be recorded by electronic data-processing systems or by photographic or other reliable means. Master formulae and detailed SOPs relating to the system in use should be available and the accuracy of the records should be checked. If documentation is handled by electronic data processing methods, only authorized persons should be able to enter or modify data in the computer, and there should be a record of changes and deletions; access should be restricted by passwords or other means and the entry of critical data should be independently checked. Batch records electronically stored should be protected by back-up transfer on magnetic tape, microfilm, paper print-outs, or other means. It is particularly important that, during the period of retention, the data are readily available.

Table 2.5. The Rules Covering Medicinal Products in the European Community Vol. IV—Good Manufacturing Practice for Medicinal Products

Citation	Requirements
4	.2 Documents should be designed, prepared, reviewed, and distributed with care. They should comply with the relevant parts of the manufacturing and marketing authorization dossiers.
	.3 Documents should be approved, signed, and dated by appropriate authorized persons.

(Continued on next page.)

(Continued from previous page.)

.4 Documents should have unambiguous contents: the title, nature, and purpose should be clearly stated. They should be laid out in an orderly fashion and be easy to check. Reproduced documents should be clear and legible. The reproduction of working documents from master documents must not allow any error to be introduced through the reproduction process.

.5 Documents should be regularly reviewed and kept up to date. When a document has been revised, a system should exist to prevent inadvertent use of the superseded version.

.6 Documents should not be handwritten; although where documents require the entry of data, these entries may be made in clear, legible, and indelible handwriting. Sufficient space should be provided for such entries.

.7 Any alteration made to the entry on a document should be signed and dated; the alteration should permit the reading of the original information. Where appropriate, the reason for the alteration should be recorded.

.8 The records should be made or completed at the time each action is taken and in such a way that all significant activities concerning the manufacture of pharmaceutical products are traceable. They should be retained for at least one year after the expiry date of the finished product.

.9 Data may be recorded by electronic data processing systems or photographic or other reliable means, but detailed procedures relating to the system in use should

(Continued on next page.)

(Continued from previous page.)

be available and the accuracy of the records checked. If documentation is handled by electronic data processing methods, only authorized persons should be able to enter or modify data in the computer and there should be a record of changes and deletions; access should be restricted by passwords or other means and the entry of critical data should be independently checked. Batch records electronically stored should be protected by back-up transfer on magnetic tape, microfilm, paper or other means. It is particularly important that the data are readily available during the period of retention.

GMP Expectations—Inspectional Citations

Firms needing to meet FDA cGMP requirements know that FDA inspectors have expectations that go significantly beyond the words written in the regulations themselves. (The FDA sets the stage for this in the opening sentence of the drug cGMPs, §210.1(a): "The regulations set forth in this part . . . contain the minimum current good manufacturing practice . . . .")

One of the more significant inspections in the pharmaceutical industry occurred in the late 1980s and crystallized the agency's expectations relating to documentation systems and recordkeeping. Selections of the FDA 483's observations are shown in Tables 2.6 and 2.7 with annotations by this book's author.

Table 2.6. FDA Form 483 Observations Related to Functional Documents (SOPs, etc.)

Verbatim Observation	*Annotation*
Failure to implement SOP revisions. There was a 3 year old revision which had become "lost" in the firm's files. Several reviewers pointed out obvious problems, but corrections were not made in reaction to reviewer recommendations. Corrections were never implemented for a number of substantive issues.	*Having a tracking system for documents in the preparation and approval process is needed.* *There needs to be a way of being sure all issues addressed during the review process are considered.*
SOPs. Responsibility for assuring the adequacy of SOP's had not been established, and virtually none of [the] corporate SOP's had documented approval by the QC unit.	*The role of the QC unit is critical in being sure SOPs (as well as specifications, lab methods) meet GMP requirements. This approval needs to be documented.*
SOP revisions. FDA evaluation of the firm's corporate and departmental revision procedures revealed the system was not formal (i.e., not organized and/or documented). The inspection revealed inappropriate or undocumented reviews, and written approvals were generally lacking. Virtually none of the corporate SOP's had been reviewed, signed, or	*The process for generating, reviewing, and approving GMP-related documents must function at **all** levels within an organization.*

(Continued on next page.)

(Continued from previous page.)

approved by the QC unit. The firm's practice was to circulate draft copies for comments without final revisions being double checked for accuracy or content.

Lack of SOP. Prior to [several months before the inspection] there was not a procedure for preparation and maintenance of corporate SOP's.

The process of preparing documents needs to be formalized in an SOP.

Default approvals. Revisions were made by document control coordinator based on comments received from respondents. The comment period was 10 working days with negative replies [comments not received] considered to be no comment (i.e., approval). . . .

No news is not necessarily good news. Even if reviewers do not have any changes, they should confirm this in writing.

Lack of [inadequate] review. Chronological file for original version . . . failed to include documented review or approvals. Circulation copies were not available, and the only evidence of review was a "Post-it" note [saying] "Looks good. Go ahead w/ minor changes you suggested". There was no documented evidence that the QC unit had reviewed or approved this SOP.

These particular observations are some of the most contentious . . . this may be an extreme view of this inspection team; rarely have other firms been hit with this observation regarding how much documentation of the review must be kept and for how long. If circulation documents are kept, it should be in a more formal way than just using Post-It® notes.

(Continued on next page.)

(Continued from previous page.)

Inadequate justifications. FDA review of recent changes revealed a failure to implement reviewer recommendations without documented justifications. . . .	*When you collect comments, you need to do something with the information. If the recommendations are not adopted, you should note why.*
[Regarding the corporate SOP for SOP revisions] Lack of approval. There was not a requirement that the QC unit review and approve final SOP's to assure accuracy and completeness	*The QC unit should be positioned in the review and approval process so they have the last word.*
Informal reviews. SOP's were established and revised by informal means. Draft documents were circulated for comments without a written definition of which personnel or departments were required to review or approve changes. Review procedures were not clearly defined (i.e., circulation was to "all concerned individuals").	*The review process needs to occur according to an SOP, including identifying those groups whose expertise and involvement is necessary. The process will move more swiftly and efficiently if the numbers of people are kept low.*

Table 2.7. FDA Form 483 Observations Related to Records and Reports

Verbatim Observation	*Annotation*
Poorly organized recordkeeping systems. Problems were found with record keeping practices across all systems inspected. Records generally failed to accurately reflect details, and the diffuse nature of available records presented a major obstacle to the current inspection. The firm's inability to readily retrieve records not only hampered the completion of the inspection, but reflected lack of assurance that management could be adequately aware of conditions and practices needed for process control. An inordinate amount of time was needed to locate information, and often there was no practical way to determine whether or not the data were complete.	*There must be a systematic way of organizing and accessing data, records, reports. The longer the inspectors are required to wait, the less faith they have in the firm's system.* *The issue of accessibility to documents is not just to please the inspector, but, more importantly, to enable the firm to use the documents to make fast decisions based on data, not just on unsupported beliefs.*
Batch records. Available records reflecting problems were not kept in the batch record files and/or were not systematically maintained in alternative locations. Instead, many of the records were found to be in various diffuse locations. For example, reports for in-process assays which failed testing were neither in the batch record files nor in the firm's analytical computer system. . . .	*There can be a variety acceptable ways for maintaining documents. The method(s) used need to be consistent and systematic, resulting in records that are rapidly accessible.*

(Continued on next page.)

(Continued from previous page.)

FDA review of the records associated with product and process failures . . . revealed they were not maintained in a formal manner. The [problem reports] were not maintained in batch ticket files and were not prepared contemporaneously at the time of occurrence by personnel performing the operations.	*GMP records need to be completed at the time the action occurs, never before or after.*
Recordkeeping. Records of [manufacturing problems] lacked standardized information and/or format. Records failed to include spaces for recording of all pertinent information	*Consistency is important. Having a standardized format for collecting information helps assure the important data elements are collected—data that can help to solve problems and monitor and improve the process.*

Electronic Records and Signatures: 21 CFR Part 11

After seven years of discussion, the FDA's regulation concerning electronic signatures and records became effective in August 1997. This regulation, crafted with input from industry, the FDA, and other governmental agencies applies not only to the pharmaceutical, medical device, and food industries, but to the FDA itself.

The document released by the FDA on March 20, 1997 (FDA, 1997), has *three* pages of regulatory text, but *68* pages of preamble—background information and explanations of what the agency intends the regulation to mean. The following is a brief summary of the regulation and does not intend to be a definitive explanation of it.

Scope—what the regulation applies to

This regulation applies to all records that are required to comply with or be submitted under any of the record requirements of the Food, Drug, and Cosmetic Act or the Public Health Service Act. In addition, whenever the agency requires handwritten signatures, electronic signatures (that meet the requirements of this regulation) will be considered equivalent, unless a special, formal exception is made by the FDA. Electronic records (that meet the requirements of this regulation) can be used instead of paper records, unless a special, formal exception is made by the FDA.

Differences between closed and open systems

The regulation requires more stringent controls related to computers systems with documents if the person(s) responsible for the documents does not control access to those documents. If it is possible "for unauthorized and potentially unknown parties" to access and potentially read, modify, or destroy documents, the system is considered an open one, and special controls must be in place to assure the integrity of the information.

For closed systems (i.e., where access is controlled) controls that need to be in place include:

- Validation of the system;
- "Ability to generate accurate and complete copies in both human readable and electronic form suit-

able for inspection, review, and copying by the agency";

- Protection of the information/records (e.g., back-up copies);
- Access limited to authorized users;
- Forced sequencing of events, if needed (i.e., you can't arbitrarily jump around in a document if you must complete it in a specified sequence);
- Authority checks to ensure appropriate persons are completing a given action or using the system;
- Data edits or verifications, if appropriate;
- Policies for accountability and to deter falsification; and
- Controls for documenting actions, including revision and change controls that provide a detailed audit trail.

Open systems are expected to have controls such as those listed above as appropriate, plus others as necessary to ensure "record authenticity, integrity, and confidentiality."

If electronic signatures are used . . .

If your system uses electronic signatures, there are several specific requirements, including the following:

- The printed name of the signer and the time and date when the signature was executed, along with the meaning of the signature, is indicated with the record, either on-screen or on the paper version;

- The signature is linked to the record and cannot be cut/pasted or otherwise transferred to another record; and
- The signature must be unique to one individual and cannot be reused, reassigned, or recycled; the firm must verify the identity of the individual before it assigns or allows someone to use an electronic signature.

Certification

Before someone uses an electronic signature, the user must certify to the FDA that they understand that electronic signatures have the same significance and legal meaning as "traditional" signatures. This certification must be done using paper and a traditional, handwritten signature. In certain cases, the FDA may request additional verification that "a specific electronic signature is the legally binding equivalent of the signer's handwritten signature."

Types of electronic signatures

There are two forms of electronic signatures discussed in the regulation. *Biometric* signatures are based upon a measurement of a physical feature of the person (e.g., voiceprint, retinal scan) or some unique, behavioral action (e.g., keystroke dynamics—the milliseconds between keystrokes and the pressure placed on each key). *Nonbiometric* signatures are similar to the cards and passwords that patrons of automatic teller machines use. Nonbiometric signatures consist of two parts: a card or log-on identification *and* a secret password.

If nonbiometric signatures are allowed, the firm needs to ensure the security and integrity of the system. This would include:

- Keeping the combination of cards/passwords/etc. unique;
- Administering the program so any attempt to use a person's electronic signature would require collusion of two or more people;
- Periodically updating or reissuing them;
- Having a loss management procedure if cards/passwords/etc. are forgotten, lost, or stolen;
- Having safeguards to prevent unauthorized use of cards/passwords/etc. and being able to detect and quickly respond if computer hackers are attempting to enter the system; and
- Initial and periodic testing of the cards/passwords/etc. to be sure they are working as intended.

If nonbiometric signatures are used in a *single, continuous period*, the regulation requires both parts of the signature to be used at the beginning, but then permits only one element to be used subsequently.

Conclusion

To stay in business, healthcare manufacturers must meet the regulations that change with technology, industry problems, and agency-driven interpretation. If a firm designs a documentation program only to meet the *basic, minimum requirements*, the company will find itself always trying to catchup to the dynamic regulatory environment.

A better, more proactive approach is to build (or revise) a system around the business, operational, quality, and regulatory issues facing your firm. Treat your documentation system as an *investment* in your operations and its people. Since almost all of the regulatory requirements have roots in quality management, when properly positioned, they will help your firm accomplish its goals. Regulatory compliance, then, becomes a side benefit.

Movable Type, 11th Century; China

Individual printing characters originated some timebetween 1041 and 1048 through the industriousness of a Chinese alchemist, Pi Sheng, who mixed clay and glue to form the movable type, then hardened each character by baking it.

Text was set by arranging the letters side by side upon an iron plate that had been coated with a resin, wax, and paper ash to grip the type. When the set plate was heated and then cooled, the type solidified in place and could be used to print multiple copies. To free the type for use in other texts, the entire block was heated until the gummy paste softened and the letters could be lifted out (Panati, 1984).

Chapter 3

The Big Picture

Goals

- Define the elements of a *documentation system*.
- Describe *functional documents*.
- Describe *records and reports*.

When your firm is developing a new drug, vaccine, or medical device, the goal isn't simply to submit the application to the regulatory agency. The goal is to have the application approved and then begin manufacturing and marketing the product. Only when the product is on the market will users benefit and all the time, energy, and money spent in developing the product be recouped.

In the same way, your organization does not implement a coordinated system for producing documents just for the sake of keeping people busy; you want the documents to help accomplish certain tasks in a controlled, consistent, cost-effective way.

Documentation System

The tasks involved in producing a document are integrated into a system. The model used here is, for simplicity's sake, a linear one with seven major activities (Figure 3.1). There are some general statements we can make about the documentation system.

The tasks do not usually take a strictly linear path. More often than not, documents are:

- reviewed,
- commented upon,
- sent back for revision,
- revised,
- commented upon,
- sent back again for revision,
- revised,
- approved,
- distributed,
- used,
- changed (i.e., updated),
- commented upon . . .

Figure 3.1. Model of a documentation system.

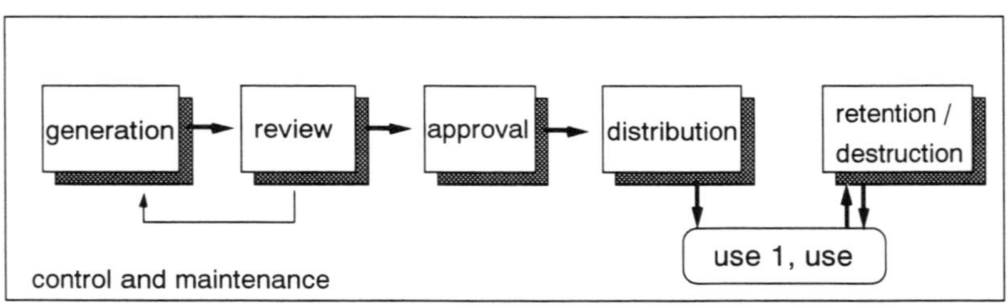

As you see, there can be many iterations and reiterations. As a documentation professional, you will want to minimize this in order to save time, money, and frustration.

The process is defined. Procedures are needed to describe the task involved in producing a document. This procedure will include who is responsible, what is to be done, and any other specifications that are important to the "customers."

"Control" surrounds the process. You can consider this as identity control for documents. Just as a firm knows something about the production, testing, and release of all the raw materials and components that go into making a product and should be able to uniquely identify each item, there should be a similar system for documents. In other words, there should be a unique identifier for every GMP document, record, form, etc. used.

This control doesn't exist just for "approved" documents; it needs to be in place as documents are being developed, so different draft versions can be kept straight without any confusion.

Maintenance of documents is considered part of the control function as it helps keep documents current with regulatory, internal, and industry expectations.

The system needs to meet *your* needs—otherwise it will not work. There is not a "one-size-fits-all" solution to documentation. The concepts and principles presented in this book need to be adapted to fit your firm's particular needs. Some of the factors that need to be considered as you establish or improve your system are:

- Regulatory requirements and expectations of agencies that inspect your firm,
- Industry practice (what the better members of the industry are doing),

- Your firm's business practices,
- Your firm's integration and use of human performance and quality concepts,
- Site and location issues,
- Money available, and
- People / skills available.

Functional Documents

Functional documents are produced using the documentation system, and, as the name states, *do* something. Functional documents are organized sets of information that:

- Prescribe—tell what to do,
- Proscribe—tell what *not* to do, or
- Define standards—specify what is acceptable and / or not acceptable.

Functional documents can be:

- Activity or task related (e.g., a cleaning procedure or laboratory method);
- Product or item related (e.g., a master manufacturing record or device master record); or
- Process or equipment related (e.g., a validation protocol).

Typically, functional documents are used repeatedly. For example a master formulae is used each time the particular item is produced; when it is copied and uniquely identified with a lot number, it becomes a batch record for a particular unit of a product.

Work instructions are another example of a functional document. These documents are more detailed than SOPs and provide very specific information for doing a particular task.

The system that produces functional documents is defined *by* a functional document. The standard operating procedure that describes the process for generating, reviewing, approving, etc. is officially produced by that system.

Records and Reports

The bottom tier of the documentation hierarchy is records and reports. They are the "containers" of data and information that:

- Communicate what happened, what was done, or the rationale behind a decision;
- Provide a historical reference;
- Demonstrate fulfillment of a regulatory requirement; and
- Provide evidence that something was done or of a fact (e.g., data).

Examples of records include:

- Batch manufacturing records,
- Packaging records,
- Sequential use logbooks for instruments, and
- Laboratory notebook.

Examples of reports include:

- Product development report,
- Technology transfer report,
- Validation report, and
- Deviation investigation report.

The format or underlying structure for a record and report is generally approved and/or described in a particular functional document. For example, an SOP on equipment calibration may require the calibration be documented on a specific form. That form would be the record—when an entry is filled in, it shows what was calibrated, who did it, when, the acceptable range, and the like.

There are differences between records and reports: Records are a collection of data while reports have a more complex purpose, such as informing management, getting a need accomplished, or, in a more defensive way, as a "memo to file."

Records and reports are prepared according to documentation standards, that is, the documents are written so entries are permanent, legible, accurate, direct, complete, clear, consistent, and truthful. Records and reports that do not have these qualities would likely be inadequate and not meet regulatory (GMP or GLP) expectations.

One of the most difficult decisions in collecting data in records is determining how much to collect. Collecting *too much* data can require time to obtain, review, and store that might not be worth the effort. On the other hand, not having enough can cause regulatory problems if you can't prove that something was done a certain way. Also, not having particular data can make problem solving difficult. Three questions that can guide you as you decide how much information to collect are:

- Why am I collecting this data?
- What am I going to do with it once I have it?
- What immediate and long-term value or risk might be associated with this data?

Conclusion

Documentation systems are among the most complex elements in your firm's quality system. They can easily disintegrate into chaos if the system is not established, managed, and operated according to good practices by people with a strong knowledge of GMP. Before going into the details of what the system includes, consider the following questions. Your answers will point to some of the areas you can focus on while reading the remaining chapters.

- Do you have a formally defined documentation system for producing functional documents such as standard operating procedures, laboratory methods, specifications, etc.?
- Are all of your documents produced by a common system?
- Has your system been thoughtfully designed and implemented, or has it simply grown to meet the immediate needs?
- Does anyone know how your system really operates?

CHAPTER 4

People and Documents

Goals

- Discuss how people use documents.
- Identify ways of making documents more useful to them.

If you are like most people, you don't read a functional document such as a procedure or specification for enjoyment. You read it to help accomplish a particular goal: you read with an intended purpose.

Think for a moment about the variety of things you read. For example, you read a novel for enjoyment. You read a movie review to learn what the film is about and what critics think of it as part of your deciding whether or not to see it.

When it comes to technical documents such as an SOP or an investigation report, you read it for particular reasons, such as:

- You need to perform the task,
- An action needs to be taken on a particular lot of product, or

- You need to be able to intelligently discuss the conclusions with a regulatory official.

There is one more variation on this: "reading to learn to do" [Redish, 1988], such as when you are reading a book like this to get an understanding of the subject so you can better accomplish a particular task.

In each of these cases, you as a reader are looking for content—information—that will help you accomplish your goal and reduce your uncertainty. The better, faster, more accurately and precisely the document communicates that content to you, the better, faster, more accurately and precisely you can reach your objectives.

Having an understanding of how people utilize a functional document or a record or report can help you design and construct documents that are more useful.

About the Users

Researchers in the fields of education and psychology have made discoveries about readers of technical information that are valuable to us as we design and produce documents.

Users are in control

No matter what a document looks like or how widely it is distributed, it is the *user* who decides if or when he or she will use it. A variety of factors contribute to the likelihood of its being used, including:

- Criticality of the goal or task,
- Impact of failure,
- The perceived value of the content,
- The difficulty or cost of obtaining the information,

- The quality of the content,
- How inviting the document is,
- Past experience with the document, and
- The ease of using the document while performing a task.

While the document producer can't modify everything on the list, he or she can have a considerable effect on many of the latter points. (Specific suggestions are found later in this chapter.)

Users approach documents differently at various times

You might first read a document, such as a validation protocol, in order to validate a process. At another time, you might use it as a model in writing a different protocol. In the course of conducting a failure investigation, you could use that same document in yet another way, as a reference. In each of these situations, your goal and the information you want to acquire from the document are slightly different. Document producers need to prepare documents that are versatile.

Users interpret as they read

Reading is an intellectually active process. You come to a document with a goal and certain expectations. As you read a paragraph, you are taking the words provided by the writer and placing them in a context. For instance, you might read a set of GMP regulations to find a solution for a particular issue you are facing. In your mind, you are probably creating a "scenario" of "who did or might do what" (Flower, Hayes, Swarts, 1983). Studies have shown that readers of regulations don't always come up with the correct scenarios, particularly if the material is written using the passive voice or with verbs made out of nouns. The result: results (i.e., interpretations) do not match those that the writers had intended.

Users rely on their background and past experiences as they interpret documents

As we obtain new information, we relate it to what we already know. We create a "scheme" that is a mental structure on which to arrange the information. This can be done by a time sequence, category, or topics. Past experiences help to shape the scheme that we build.

There can be dangers in using past experiences to interpret new ones. The previous experiences might not be a valid match to the current situation. Or, someone might make a premature match that isn't correct.

Users employ documents as tools

Technical documents are useful when they help people accomplish a goal or task. They are one of many resources that a worker has access to. Like other powerful tools, documents must match the need and efficiently assist the user. A hammer is not the best tool in all situations; a narrative paragraph is not the best format in providing information in all situations.

Creating Documents that Effectively Communicate Content—General Comments

Knowing how people use documents is the first step in creating ones that better communicate information. Below are some ideas that can make documents better communication tools. Many of the points are questions you need to ask in your particular situation, with the answers shaping the documents you produce.

Understand the audiences

- Who are the users—scientists, operators, technical experts, managers?
- What are the reading skills of the users in the language of choice?
- Where will they be using the documents—in the lab, manufacturing areas, or office?
- How much do they know about the process or activity?
- Might regulatory agency personnel also review these documents with a different level of understanding of your operations?
- What questions might these people have?
- Where will users go if they need more information?

What this means: Your answers to these questions will help define both the content and amount of detail included in the documents and also the level of difficulty at which the information is presented.

Help make connections with the user's expectations and experiences

- What are the goals of the task and steps?
- Are there "generic" terms or steps that are common?
- What are typical questions that people have asked about this?

What this means: You may find it useful to arrange information based on the goal or the outcome (e.g.,

"Investigating Deviations Using SOP 1314" or "Reviewing Clinical Submission Forms for IRB Signoffs"). Invoking a "theme and variations" approach can be valuable if there is one basic flow used for accomplishing a task with different variants. This way, readers will get the "theme" and see that slight changes can be employed in certain situations.

Keep things simple

- Are there standard ways you can present information to the users?
- Is there a way that people can easily and quickly find particular information?

What this means: Establish and consistently use a set of formats that communicate the content. This does not mean that one format is used for all procedures; it does mean that if three formats are used, they are designed to meet particular needs. For example, you may have a decision tree format used in certain procedures and a three-column format for operational procedures. Having a specific location on a document for titles, control numbers, etc. can be very helpful to users.

Give readers a context

- What is a one- or two-sentence description of what people are needing to do and why?
- Is there a way for the user to quickly confirm they are going down the correct path?

What this means: Provide a procedural introduction to the task before listing the steps required to perform it. Structure the information so readers see the goal up front, not at the end, allowing them to confirm that this is really what they want to do (Figure 4.1).

Figure 4.1. Example of using a goal in the beginning of a task sequence.

I. PURPOSE . . .

II. SCOPE . . .

III. TASK OVERVIEW

Correctly labeled clinical materials are critical for patient safety, the proper administration of the study, and regulatory compliance. Regulatory Affairs personnel perform the final check of the label copy before the labels are placed on the containers and sent out. The Regulatory Affairs review assures that all federally required label elements are present and that the information is correct and not misleading.

IV. PROCEDURE

If the label copy meets the criteria below and is considered satisfactory,

then the labels can be approved for use.

Otherwise, deficiencies are communicated to the clinical representative who provided it.

A. Inspect the copy of the label being reviewed.

1. The label meets the general criteria for any regulatory document. The information provided is . . .

Be consistent

- What is the preferred way of expressing units of measure, spelling a particular word, identifying equipment, etc.?
- Does your firm use words ("jargon") that are unique to it?

What this means: Adopt and standardize particular choices, and communicate these by way of a style sheet. Establish a company-wide glossary with definitions of words that are unique to your firm or that have a particular meaning in your organization (Figure 4.2).

Use typefaces and layouts that promote readability

- Where will the documents be used?
- How will the documents be accessed (on paper? using a computer monitor?)?

Figure 4.2. Example taken from a corporate glossary.

BOM	Bill of Materials.
BPS	Bits per Second.
CAD	Computer-Aided Design.
CDROM	Literally, Compact Disk Read Only Memory. This term describes the use of the compact disk to store information that does not (and cannot) change.
CFR	Code of Federal Regulations. A compilation of all final Federal regulations that have been published in the *Federal Register*.

(Continued on next page.)

(Continued from previous page.)

cGMP	Current Good Manufacturing Practice. The U.S. FDA's current regulations covering drug manufacturing–21 CFR 210-211.
CPA	Capital Project Authorization.
CPU	Central Processing Unit.
DCR	Document Change Request. The form used to request, approve, or effect document changes.
EC	European Community (formerly known as EEC).
FDA	The U.S. Food and Drug Administration.
gmp	The generic concepts related to good manufacturing practice.
GMP	Good Manufacturing Practice. The regulations of all groups and countries that address the topics manufacturing drug products for use in people and animals.
HPB	The Canadian Health Protection Branch.

Conclusion

"User-friendly" is an overused term with a lot of cynical implications. When it comes to the documents you produce, you want them to more than friendly. They need to be inviting, helpful, and structured to be of assistance to people as they do their job. Take a sampling of some of your documents and document users and ask them the following questions:

- Do the documents you or your firm produce *help* people accomplish their goals?
- How "user friendly" would users rate the documents?
- Have you asked document users if there are ways to improve the usability of the documents?
- Do your documents serve the needs of the writers/generators? Routine users? Infrequent users? Management? *Everyone*?

Dictionary, 600 B.C.; Mesopotamia

The earliest preserved dictionary is an Akkadian word list from central Mesopotamia dating from 600 B.C. About this time the Western-style dictionary was emerging from Greek philosophers, who had begun to analyze speech patterns and language, establishing the roots of grammar and syntax. The Greek language already was in transition from its archaic origins to its classical beginnings, and over the next several centuries it continued to evolve to such an extent that by the 1st century A.D. commentaries on words and their usage were an absolute necessity. After a lengthy 1st-century lexicon by Pamphilus of Alexandria, the Greeks became skilled compilers of dictionaries, which they revised frequently (Panati, 1984).

CHAPTER 5

Documentation Systems: Control

Goals

- Define *control* in relation to documentation systems.
- Discuss various types of controls imposed on documents.
- Differentiate between "dumb" and "intelligent" numbering systems.
- Identify and discuss issues related to the control of documents.

Before you produce your first document, you need to consider how you will control them. Control has several aspects, including:

- Identifying each particular document,
- Identifying the versions or "releases" of a document,
- Identifying related documents,
- Identifying each page of a document,
- Assuring that only authorized people make changes to documents, and

- Preventing the release/use of a document before its effective date.

Distributing the documents also needs to be controlled; this is discussed in detail in Chapter 10.

General Comments About "Control"

Controlling documents is one of the most critical aspects of a documentation system. Control is necessary to keep documents in synchronization, meaning that all users, at a given time, are working with the same, exact version. Having a document system that is controlled prevents unauthorized people from making changes—intentional or unintentional—in documents. Control needs to be integrated into the very fabric of a documentation system.

Each activity involved in producing a document (e.g., generation, review, etc.) has some particular issues related to control; these will be discussed in the chapters where the activity is described.

One other aspect of control is *change control*, also known as *configuration management*. Change control is a mechanism for evaluating the technical, quality, and regulatory appropriateness and ramifications of a change. While change control needs to be connected to control, it is a separate issue; it will be discussed in more detail in Chapter 17.

Document Identification

Just as a batch of tablets or an individual medical device is uniquely identified, each document you produce must have an identification.

Your documents will have at least two identifiers:

Title. A set of words that describe the document's contents. For example, for an SOP, "Cleaning the Stokes Tablet Press Model 101-T," or "Bill of Materials for Producing Pacemaker Model 15483-A." Titles need to be short, descriptive, and unique so a user looking for document can easily find it.

Document number. A set of numbers/characters that uniquely identify the document. They can be simply a sequential number assigned from a log book or have some "intelligence" behind them. Depending on how your system has been set up, the document number may or may not include the version or "rev level" of the document. The document and version numbers together identify a particular document.

You may also have another identifier useful to those preparing and maintaining the document: the file name. This is a set of characters that point to the electronic file that contains the document. The exact file name depends on the word processing or document management system your firm uses, with some systems (e.g., Macintosh) allowing for longer, more meaningful names.

A Closer Look at the Document Title

Besides being concise and unique, titles need to be meaningful to the users. Key words, for example, "cleaning" or "releasing," should be present that give the reader a clue to when this document would be used. The title should give the user specific information, such as the name of an instrument or product. It is not necessary to begin a title with "SOP for. . ." or "Operational Method to. . . ."

A Closer Look at the Document Number

All document numbering methods can be broken down into two basic systems: "dumb" and "intelligent."

An example of a dumb numbering system is sequentially assigning a number to a document. For example, your first procedure is number 000001, your next one is 000002, etc. The only information you can derive from the number is that one document was assigned a number before the other (and not necessarily that one document was approved or issued before the other).

Intelligent (or "smart") numbering systems are formatted in specific ways. If you understand the format and codes used in the identification number, you can gain additional knowledge about the document. For example, you may have a system with the following format:

$$$-####-@

where

> $$$ indicates the type of document, such as procedure, specification, drawing, or work instructions;
>
> #### indicates the identity number of the document; and
>
> @ indicates the version or "rev level" of the document.

In practice, this could look like: SOP-0254-B and mean that this document was the second revision of the SOP number 0254.

Usually, intelligent systems are arranged so that the information that changes most frequently or is most specific is on the right of the set of characters.

Sometimes intelligent systems can become extremely complex, for example, having seven or eight bits of information within a format. An advantage in using an intelligent system is that it can help you better organize and access information. In scanning through a list or doing an electronic "sort," you can easily find particular sets of documents.

On the other hand, intelligent systems have distinct disadvantages. First, they need to be carefully planned in advance, considering both current and future needs. One firm seriously underestimated the number of procedures that they would have and incurred a large expense in changing from a three-digit format to four.

To work properly, intelligent systems have to be tightly controlled. Everyone involved, from local departments to remote sites, needs to agree on the codes. Any change, such as local additions, will affect the integrity of the entire system. Intelligent systems need to be kept simple. Each element added to a format should be evaluated to determine if it really adds value and does not cause complications.

Which should you use? If your system is electronically based or if you keep your document index on a database, there is not a great deal of need for a complex numbering scheme—a sequential number will work very well. If your organization has the culture that permits a strong, centralized operation for controlling numbers, and you have a paper-based system, a numbering scheme with some intelligence can be used.

Identifying Revisions to a Document

The title and document number identify the topic covered by a document. The revision or version code identifies the specific content in a document. The revision code is required, along with either the title or document number, to pinpoint a particular document.

Revisions codes are sometimes called "editions," "revs," "rev levels," "releases," or "versions." In any event, the code allows you to differentiate between documents that have been changed in some way. Sometimes the revision code is built into the document number by way of a suffix. For example, 9876.2 or 9876-B both refer to the second version of the document. Sometimes you may see a version indicated by "rev 2."

Software manufacturers identify their versions or releases in several levels. For instance, the Macintosh Operating System on a computer could be 7.5.3L. Such a code may be useful to identify a complex program, but would probably be too much for a specification or other document in a pharmaceutical facility. A simpler way, such as increasing the release number by one digit or letter each time it is released, would be less confusing to everyone.

Identifying Documents that Relate to One Another

Frequently a document refers to another document. For example, a cleaning procedure specific to a stainless steel tank may refer to preparation of a detergent according to another standard operating procedure. It is important to know about the references in order to make or evaluate changes.

One of the simplest ways of doing this is to list documents referred to in the document. For example, at the beginning or end of a validation protocol, include a section entitled "Documents referred to" and then list the identification number. Unless there is a specific reason, do not include the revision number, as including it (e.g., SOP 123, Revision #4) means that the specified revision of that document is the only one to be used. If just the document number is given, it will imply that the current, approved version is the appropriate one. (Be sure to define this in an SOP and include this as a topic to cover as you train SOP users.)

Another way is to build a database that links related documents. This is a standard feature in some document management software applications; alternatively, it could be developed using a simple spreadsheet or database application. One medical device firm created their own database similar to their bill of materials system. They could query their system to determine which procedure was used in other procedures.

Identifying and Controlling Pages

Each page of a document should be positively identified with the document number and version as well as the page number and total number of pages. This way, if a page becomes detached from the document, it can easily be identified.

For example, you may have a format with a header or footer on each page with

"SOP 9876.2   page 5 of 7."

Attachments, such as flowcharts or appendices should be included in the page count.

Controlling Changes to a Document

Control is not just identifying a document and changes made to it, but limiting who can make and publish those changes. This type of control needs to be defined by procedure and incorporated into any word processing systems that are used.

Defining security levels (e.g., read only, read/write, read/write/replace) must be done with any documents that are available electronically. Most basic systems, however, allow users to obtain a document and copy it into a different file and then modify it. More sophisticated electronic systems make this difficult.

Some of this control deals with document distribution; other aspects need to be incorporated into the distribution activities so users can easily obtain the approved, current version of the document.

Special attention must be given to the potential of deleting a document, records, and reports. Deletion is not done of GMP records; modification can be permitted, but as shown in the GMP requirements in Chapter 2, agencies expect that the initial entry, even if in error, be legible and the reason for the change to be included as necessary. Simply deleting a document before its official destruction date (see Chapter 11) is improper practice.

Preventing the Release/Use of a Document Before Its Effective Date

Before documents are approved for use, they are generally circulated for review and approval. While most people would not knowingly use an unapproved document, you need to take steps to prevent unintentional use or confusion. Part of the control of this would be defined in your procedure for document distribution to training and other implementation personnel.

Stamping a document with "For Review and Approval Purposes Only" or printing such copies on distinctive paper can help. Also, placing an electronic water mark across a page will also make all readers aware of the document's approval status.

Conclusion

Control needs to be totally integrated into your documentation system. Every action and every document should be appropriately controlled. Creating a simple system is easier said than done, but thoughtful planning is the key. Another piece of maintaining control of documents is to have the system meet the needs of the users so that they do not bypass the system. They need to be convinced that well-identified, properly released documents will allow them to work more confidently.

Consider your own current system. How well does it match up to these expectations?

- Each document we prepare is uniquely identified with a meaningful title, number, and revision code.
- The numbering system we use to identify a document is well defined and controlled.
- If a page from a document becomes detached from the other pages, we can easily identify the document that it came from.
- It is easy to identify an unapproved document.
- It would be extremely difficult, if not impossible, for an unauthorized person to modify an approved document.
- Our process of controlling documents is clearly defined in a procedure.

Frequently Asked Questions

Q. *I'm releasing the first version of a specification; it has not existed before. We indicate revision number by a number after the document identifier (e.g., Spec 3455-1). Should this first document be "-0" or "-1"?*

A. You could do either, as long as your practice is defined (and followed) in your documentation procedure. Typically, it would be your first edition and would be called "-1".

Q. *We are circulating SOPs for approval. How do we keep each draft version identified? We don't want to change the revision number until the document has been formally approved for use.*

A. A simple way would be just to use the document's identification and revision number with a supplemental designation of DRAFT #. This way, reviewers know the document that was the basis for the draft.

Chapter 6

Document Generation

Goals

- Identify the three main tasks of generating documents.
- Discuss the steps involved in each task and what is produced.
- Discuss who the writers can be.
- Differentiate between and discuss the implications of new ("first-time" functional documents) and revised documents.
- Discuss controls applied to the generation phase.

While generation is the start of the process of creating documents, there are some that need to exist before you start document generation . . .

- Have an SOP on producing documents that provides details on the generation task, as well as on document review and approval.
- Have controls (e.g., numbering, revision) in place. (See Chapter 5.)

- Have styles and formats defined. (See Chapters 4, 13, and 14.)

Preparing functional documents, be they procedures, specifications, laboratory methods, or manufacturing instructions, means more than simply sitting down and writing them. They must be carefully planned so they meet user and organizational requirements. Data must be collected and organized and then converted into the desired document according to standard styles that make it easy to use the document.

Task One: Planning

Planning a document is done on two different levels. First, it is done when you create the style to be used for all similar documents, for instance, all specifications or laboratory methods. Here you want to have a consistent look for all of your documents, so users can easily find the information that they are looking for.

Second, you will plan each individual document, such as a procedure, to ensure it contains information important to the document users and it is at a level of detail they need. Your focus here is on the content and clarity of presentation. Some writers call this step *pre-writing*.

As you define each type of document, and again as the individual ones are written, it is critical to know who will be using it. This is called an *audience analysis*. Answers to questions like

- Who are the users?
- How will they be using the information?
- What other tools or resources are available to them?

- How much experience do they have?
- How often will they be using this document?
- How will people be trained on this document?

will influence how much detail is included.

For example, if a laboratory method is being generated for a relatively simple testing activity that is being done every day by well-trained, experienced experts, you may not need to include as many of the details—key steps and substeps may be sufficient. On the other hand, if the method is going to be used by many analysts in laboratories around the world who may or may not have years of experience, you may want to include more details, particularly the "art" or "tricks" that successful performers have learned to use.

Once an audience analysis is conducted on a group of users, it does not need to be repeated unless there are significant changes to the group.

Task Two: Collecting and Organizing Data

This task is sometimes combined with the task of actually writing the document; however, writing goes much more smoothly if information is gathered and arranged in advance. The time required for this step varies depending on the type of document that is being produced, the knowledge that the preparer has or has access to, and the depth of detail that is required.

Before starting to collect the data from subject matter experts or other sources (e.g., development reports, validation protocols), it is important that the collection be planned. This saves you and those you are working with time. For example, if you are developing a specification for rubber stoppers that will be used with a parenteral product, you might want to prepare a list of questions that can answered by reading the development report or talking with experts. Your questions might include:

- Composition of rubber required?
- Size of stopper?
- Color?
- Size of container that stopper will fit in?
- Drawings available of the complete container/closure system?
- Machine/process capability?
- Proposed processing steps? (e.g., washing? sterilization? siliconing?)
- Proposed lot size/stoppers needed per run?
- Experience of development and manufacturing personnel?
- Regulatory, compliance, or compendial requirements?

If you are writing a procedure defining a process (i.e., task), you would collect data on:

- The information and materials that are the inputs to the process,
- The information and materials that are the outputs of the process,
- The "actors" involved in the process and what they do,
- The steps and substeps that occur in the process,
- Steps that require verification, and
- Decisions that are made.

Chapter 13 provides a template for collecting information that is used in procedures and methods. Other sources for documents may come from technical publications or presentations at conferences. Trade publications like *Pharmaceutical Technology*, *Pharmaceutical Engineering*, and *Medical Device and Diagnostic Industry* are good, practical sources on what should be included in a validation protocol, master validation plan, etc.

As you organize the data, think ahead to how it will be most clearly presented in the final document; you may be able to reuse much of it.

Task Three: Writing the Document

With the data you've collected and organized, you are now able to write the functional document. Guiding you in its preparation are the standard styles and formats that your firm has selected and defined.

Also shaping how you present the information in the document is your knowledge on who will be using it, gathered from the audience analysis. Documents are used by different audiences for different purposes at different times. For example, raw material specifications are shared by your purchasing department with vendors in the bidding process. Specifications are also used by your laboratories in conducting sampling and material testing. Your quality control unit uses specifications as the basis for deciding whether a given lot of raw material is acceptable. Other groups, such as validation, would use the specification as a reference.

The styles for presenting the information in a functional document need to help people use the document easily in a variety of situations. (More details about styles and writing documents are found in the following chapter.)

Who is the writer?

There is a variety of approaches that can be taken in generating the document. The key, however, is obtaining the content from those who know it the best. The following examples are two ways drug and device firms are actually writing documents.

- **Subject matter experts as writers.** One firm provided technical writing courses to their operational staff so they could write procedures they and their colleagues would then use.
- **Subject matter experts as contributors; staff professionals as writers.** Another firm hired technical writers to work with their engineers who provided the details that the writers would prepare into documents.

One approach that should be avoided is that of using writers who are removed from the details of the activity. Experience has shown that people in development or technical service groups may have a great deal of expertise, but little knowledge of how things are done in an actual work/production area, such as formulations. A solution may be to have these experts collaborate with operations or other line personnel.

Initial Documents, Revised Documents

There are different challenges in creating the first version of a document from those in revising one to reflect changes.

The initial document is prepared because of a new process (e.g., needing specifications for new components) or when a gap is observed (e.g., a task that was never formalized before). In these cases, there has not been a functional document in place. Any document you write in this situation will either be validated or will define an existing (and, you hope, validated) situation.

For example, preparing a specification for packaging components will be based on past stability/compatibility experience and will be confirmed by additional stability testing with the specific drug product. Another example is writing a procedure that describes how someone has been making a cleaning solution—the operator has been preparing it and the SOP formalizes what has been happening. A third example is a laboratory method for a new clinical trial material that is starting its Phase I study; in keeping with current GMP expectations, the method will soon be validated.

Because these are "baseline" situations—you are simply defining what currently "is"—these documents do not need to go through a change control process. They must, however, meet all the appropriate regulatory requirements and expectations.

Revised documents are handled differently. They *may* need to go through a change control process. In your firm's change control procedure, you would define what a change is, how it is evaluated, and what groups are involved in the approval. (See Chapter 17.)

If there is a change in the content, the proposed change must go through the change control process. For example, for a procedure decreasing the frequency of in-process control testing from once each half-hour to once an hour committee members would consider how this change meets corporate policy and regulatory commitments, as well as any validation implications and process ramifications.

Controls During Document Generation

Until documents are actually approved and made effective, they cannot be used. Your firm should have controls in place that will make it difficult, if not impossible, for people to obtain documents that are in the development phase. This is especially important if significant changes are being made to the process or components.

Some ways of preventing unauthorized use of documents that are being developed include:

- Restricting access to electronic copies of the document;
- Minimizing the number of copies being printed and used;
- If copies are needed, marking them with highly visible stamps, such as "Do Not Use: Document in Development" or printing them on specially colored paper; and
- Destroying all draft copies in a shredder after they have been used.

Conclusion

Those involved in implementing electronic document management systems say that the largest amount of time spent creating and publishing documents is in the generation phase. The time and care that is invested here in crafting an accurate, precise SOP, method, drawing, etc. will have a payoff in that the reviews, approvals, and implementation will go faster.

- For each type of functional document you are developing, you have identified the users and their document-related needs.
- You have defined the approximate level of detail that should be contained in a given set of functional documents.
- The writer of the document has the needed skills and access to information sources and subject matter experts.
- As the writer prepares to develop a particular document, he or she has collected data on the topic from knowledgeable sources and subject matter experts.
- The writer has organized the data in a way to help understand the content before he or she begins to write the particular document.
- The writer has selected the approved style that is best suited for presenting the information.
- If the functional document is a new one, you have considered the regulatory and validation implications.
- If the functional document is a revision of an existing one, you have evaluated it in terms of your change control process.
- "In-process" documents are physically and/or electronically controlled so they would not be inadvertently used.

Spelling Mistakes

Spelling mistakes are not very common in adult handwritten texts—descriptive studies suggest that, on average, about 1% of letters are affected, and 1–1.5% of words. One study classified errors into four formal types: omissions (buton), additions (hopefull), substitutions (attendence), and inversions (tabel). Other possibilities are rare. There is very little difference in the first letter. Insertion errors are much more likely towards the end of a word, whereas the other types are more likely towards the beginning. Substitutions and omissions seem to go together (Crystal, 1987).

Quill Pens, 6th Century B.C.

For more than a thousand years, the quill was the standard writing instrument for people of many civilizations. Birds, of course, supplied the basic instrument, and man adapted it to his needs. First a freshly plucked (or found) quill was dried, then cleaned of all oily and fatty materials that would interfere with the absorption of ink. To stave off brittleness, the quill was dried by the slow application of gentle heat.

Large wing feathers made the best quill pens, and three fowl offered the finest choices: swans, turkeys, and geese. Geese feathers were preferred because of their sturdiness (although some of the finest early etchings and writing apparently were done with elegant black crow quills) (Panati, 1984).

Chapter 7

Preparing Useful Documents

Goals

- Describe the considerations that go into producing functional documents.
- Distinguish between decisions made initially and those made each time when preparing a functional document.
- Identify and define the different parts of a document.
- Show different ways to present information.
- Establish and use templates, styles, and formats.

As you create different sets of functional documents, you are creating individual structures similar to buildings and houses, as well as a community of these units.

There is something to be said about the spontaneity that exists in towns that spring up without any planning or zoning, but there are difficulties as well. Having a master plan for a community takes up-front time and effort, but the results are much more satisfactory to all of the residents.

Having a "master plan" for your documents—what they will look like and contain—is extremely useful. Consistency will give your documents a more professional look and help users work with them more efficiently. Once you have identified the different functional documents your organization is going to use and understand the relationship between them, you can begin designing particular types of documents.

Whether you are an architect or document designer, you are trying to balance a number of different issues, such as esthetics, functionality, laws (natural ones like gravity as well as man-made such as zoning), "flexibility" for future modifications, final use, materials of construction, resources, and environment.

This chapter is concerned with what documents look like—their appearance and underlying structure. What goes inside the document—its content and how it should be aimed at the users, is covered in Chapter 4.

Models, Templates, Formats, and Styles

There are several terms that are used related to documents that are sometimes misunderstood and frequently misused. Four particular ones are discussed below.

A *model* is a completed document that includes content. It may be an actual "working" document or one that is specially created to show how various styles are used. For example, Figure 13.2 is a model of a policy.

The *template* is the underlying structure of a document. It illustrates how the content is to be presented (e.g., paper or on-line) and include sections and styles that are used for a particular set of documents. Figure

7.1 is an example of template with sections that are annotated to better help the users.

Figure 7.1. Example of a Template.

Corporate Standard Operating Procedure
Acme Pharmaceuticals

Procedure Title

(Use a unique, brief, descriptive title that relates to the process and/or the action accomplished)

SOP No.: ____

Revision date: ____

Effective date: ____

Validation No.: ____

Purpose	*Describe the task that this procedure covers and what is accomplished or produced when the procedure is successfully completed.*
Scope	*Describe the sites, areas, plants, and/or departments that this procedure applies to.*
Reason for revision	*Describe the general reasons why, if this is a revised procedure, it is being re-issued. If this is a new procedure, insert the words, FIRST VERSION, in this space.* *As needed, mark the beginning of sections that have changed with a* **Δ** *symbol.*
Definitions	***Word to be defined.*** *Define words that are important for the correct implementation of this procedure. Also define words that have a specific meaning in this SOP's context.*
Tools, equipment, needed	***Tool name.*** *Include documents; other SOPs; safety equipment; and other information, things, or people who are needed to safely and efficiently perform the steps.*
Initial situation	*Include description and drawings of how the user will commonly find the situation, e.g., labeling machine that has just completed a product run.*

(Continued on next page.)

(Continued from previous page.)

General requirements	*Discuss the general requirements for using this procedure or other items that are important for the user to know or do. Provide the information in a question-and-answer format.*
Time limitations	*This is a particular requirement that must be emphasized in this procedure.*
When is it started? *When must it be completed?* *What if ...?*	*Again, typical questions are posed and answered, based on a knowledge of the users.*

Procedural content

Step	Who does	Action taken
1.	*This is the person(s) who is involved. Titles are used.*	*This is what the person (listed on the left) does. Each action starts out with an action verb.*
2.	*As other people are involved, they are included.*	
3.		
n . . .		

Attachments	**Attachment name.** *(List all of items that are attached.)*
References	*Other procedures referred to in this procedure, corporate policies, and primary documents are named in this section as well as where they can be obtained.*

(Continued on next page.)

(Continued from previous page.)

Approvals *This section is done in keeping with the SOP regarding SOP approval.*

This procedure has been approved for use by:

Director Quality Control	(name)	(date)
Director Quality Assurance	(name)	(date)
Director Manufacturing Ops.	(name)	(date)
Director Regulatory Affairs	(name)	(date)

++

Prepared by and date:
Released by and date:

File name

A *style* is a set of formatting instructions, such as typeface and size, indentations, margins, etc. For ease of use, as well as appearance reasons, styles should be limited in number. A style sheet is the listing and definition (much like a specification) of all styles used in a document.

Formats are specific definitions that are applied at the character level (e.g., typeface and size) or at the paragraph level (e.g., space between lines, borders around paragraphs).

Creating a Template

Developing a template can be as easy as modifying one that already exists to meet your particular needs or as complex as creating one from scratch.

If you are creating a template, the typical practice is to use a "grid" and arrange all the sections. The grid helps to show the relationship between the page, margins, headings, subheadings, footers, page numbers, etc. As you create or modify a template, there are things you need to consider, including:

- **Page size**. There are different "standard" sizes around the world. If a document is to be shared in various countries, you will have to think about printing and photocopying so all information (especially headers and footers) is present. This is also important in electronic systems where screen size and application software play a major role.
- **Binding.** How will the pages of the document be assembled? How will the documents be maintained? Simply stapling pages together might work for infrequently used procedures, but might not be rugged enough for operational procedures that you expect people to use in a manufacturing area. The type of binding you select will affect the margin size required.
- **Page layout**. The two principle layouts are "portrait" and "landscape." Usually documents are printed out in the portrait position; logs and other data-capturing documents are often printed in the landscape mode.

For some documents, for instance, handbooks or glossaries, you may want them to be an odd size so people can more easily find them on their desks or keep them in their pockets.

Reuse of elements

Unless there are special reasons for not doing so, try to use and reuse different document elements whenever possible. This saves time in designing the document, but it also helps achieve consistency between the documents and reduce confusion. For example, if you have a signature block on the bottom of a procedure, lab method, drawing, specification, etc. use the same one for all documents. Electronic document management systems or those applications that consider a document to be a collection of objects make this use and reuse simpler.

Sections

As you define what is to be contained in a particular document, you can easily define the sections the document is to include. It is the content and how it is most logically presented that influence what sections and their arrangement. Hierarchies are very important as you relate a section to a subsection to a sub-subsection.

Styles

A section can consist of one or more styles, i.e., ways the content is presented. The most important aspect of selecting a style is to find one that clearly and completely communicates. Styles need to guide the user, not confuse or distract. Consistency is very important in applying styles. The fewer choices that are available, the most consistent your documents will be.

Formats

A given set of individual formats comprises a style and is chosen as much by user considerations as by personal/organizational preferences and how the documents will be used. For example, type size in functional documents is usually 11-12 point; any smaller can make reading difficult, especially in manufacturing or laboratory setting.

A Look at the Research

Colin Wheildon, in his book *Type & Layout* (1995), did one of the first in-depth scientific studies on how people use printed information and the effectiveness of the styles, formats, and typefaces, particularly in print-based advertisements. Many of his findings are applicable to other documentation endeavors as well. Some of the more interesting points he makes include:

- 65 percent of readers showed poor comprehension of articles set in sans-serif body type;
- 40 percent showed poor comprehension of articles set with margins ragged on the right-hand side, and 72 percent showed poor comprehension of articles set with ragged margins on the left-hand side;
- The darker the headline, the more understandable it is;
- Black text printed on a light tint is very comprehensible;
- Headlines and text set in all capital letters are almost twice as difficult to understand as those set in mixed-cases;
- Most readers find subheads useful, especially in longer articles;
- Both wide measure (i.e., text across a page) and narrow measure (thin columns of text) are hard to read; and
- Readers are easily annoyed; most complained about long, wordy headlines.

Wheildon summarizes his work with six points, "How to Drive Readers Away":

1. Set the body copy (e.g., text) in a sans-serif type. Use the smallest typeface you can find.
2. Set some of the copy so that it has a ragged left margin.
3. Set other copy so it has a ragged right margin.
4. Stretch lines of type all the way across the page.
5. For variety, use a multicolumn format and include some very narrow columns
6. Use a difficult-to-read typeface, and then, underscore an entire paragraph.

Style Guides

When a prospective author is considering submitting an article or manuscript for publication, he or she would follow a style guide provided by the publisher. This gives the author all the information needed to produce a document that meets the publisher's specifications.

Firms also can develop their own style guides. Typically these would include the "official" format for writing letters, internal memos, e-mail, press releases, and the like. Style guides can also help define GMP and regulatory documents, like submission cover letters, recall letters, "Dear Doctor" letters, and similar documents.

There are other style guides that are helpful in preparing documents, such those from the University of Chicago Press, and the New York *Times*.

One newer style guide is produced by Hardwired Press, affiliated with WIRED magazine. The information it contains is very useful—acronyms, rules for hyphening—but a little difficult to access.

Conclusion

As you are preparing your templates, styles, and formats, they need to look good and appealing to the user, but also meet the reader's expectations and needs. When it comes to design of documents, simpler is better.

Frequently Asked Questions

Q. *Everybody in our firm uses a slight variation when referring to electronic mail. What is correct—email, E-Mail, e-mail ?*

A. This is the type of question that would be defined in your firm's style guide, as practice differs. For example, *WIRED Style* (1997) prefers the term "email" while the publisher of this book prefers the hyphenated form "e-mail."

Q. *How is it best to indicate an e-mail address in a document?*

A. Deferring to *WIRED Style*, it is to be set in all italics, such as *info@learningplus.com* Some people like to put angle brackets around the address, as in *<info@learningplus.com>*. Be very careful where the breaks appear; if the address needs to be broken, do it just before the punctuation, pulling the period or @ sign to the next line.

Chapter 8

Document Review

Goals

- Define *document review.*
- Describe two stages of document review.
- Identify who is involved in document review and what they do.
- Define and discuss approaches to routing documents for review.
- Describe documents that are generated and kept during the review phase.
- Identify and discuss control issues related to document review.
- Discuss alternatives in reviewing documents.
- Identify issues related to document review and discuss ways they can be overcome.

With the first draft of your document completed, you are now ready to begin the review phase. Alternatively, document review occurs when documents have been revised based on comments made during a previous review cycle.

Document review is defined as the formal process of having knowledgeable individuals read and comment upon the appearance and content of a particular document from multiple points of view. The comments are then provided to the document originator, who evaluates the comments and changes the document as appropriate.

The end result of a document review is that the document is prepared correctly according to standard formats and styles, that all information in the document is correct, and that the document can be successfully used for its intended purpose.

Document Inspection

There are two stages to reviewing a draft document: *document inspection* and *technical review*.

Document inspection is examining a draft document for its conformance to the standard formats and styles that the organization has adopted. In addition, inspection takes into account the mechanical and grammatical conventions that should have been followed in generating the draft document. Sometimes, inspection is also called *verification*.

Document inspection is the first stage of the review process. In some organizations, it is thought of as the last step in the generation phase. Where it is placed is not as important as the fact that it is done early in the review process.

Persons who inspect the document do not need to be content experts. Rather, they need to have a thorough understanding of the formats and styles used in the firm as well as a solid knowledge of grammar. Technical writers or editors are very useful people to have inspecting the document.

Document inspectors are mostly concerned with *how* something is said and the method used to present the information; the content will be examined in the technical review stage. Table 8.1 presents some examples of what a reviewer would look at during the inspection process.

Inspection has been made easier with some of the tools that are integrated in word processing applications or available as add-ons. "Spell-checkers" are one of the simplest of these tools; grammatical checkers or expert systems are also becoming more prevalent and reliable for English, French, and other languages.

While these tools help, they need to be used with caution. Almost everyone can tell an embarrassing story or two of when the wrong word was used, but spelled correctly.

Like other forms of inspection, examining a document is best done by someone who was not involved in its preparation. This brings a fresh set of eyes to the task. If there is not a second person who can help, the writer can be more effective in reviewing the document if some time has elapsed between when it was written and reviewed—even an hour or two can help.

Jonas and His Notebook

Jonas Salk spent 98 percent of his time documenting the things that didn't work until he found the thing that did (Wurman, 1989).

Table 8.1. Document Attributes Examined During Document Inspection

- Spelling
- Grammar
- Hyphenation
- Use of standard units of measure
- Use of standard abbreviations
- Proper labeling and formatting of figures, tables, appendices
- Use of approved document formats
- Logical organization of the material
- Appropriate application of headings, subheadings
- Correct use of numbering schemes, bullets, indentation
- Consistency with other similar documents
- Presence of front/back matter
- Use of correct formats for phone numbers, e-mail addresses
- Accuracy of standard information, e.g., address, phone numbers

Technical Review

The technical review of the document examines the content from various points of view using people with expertise in the particular area. This can include people who are experts in the technical details being presented as well as those from Regulatory Affairs, Quality Control, Validation, Compliance, etc.

There are two major elements to the technical review. First, the review considers the specific content: that it is technically correct and can be done as written, and the level of detail is appropriate to the users of the document.

Second, the review considers the relationship of the content to other documents. For example, if reference is made to other procedures, are they the correct ones; if cleaning logs to be created as part of a cleaning procedure, do the logs include what the GMPs require? The technical review is based on the concept that not one person in an organization is an expert on everything, and therefore using a "team" approach can provide multiple perspectives on a document.

There are various methods that can be used to review the technical content of the document; the one(s) selected depend in large part on the type of document reviewed. Table 8.2 lists different methods in order of their complexity.

Some documentation specialists who do not come from a medical device or pharmaceutical background use the term "validation" in describing this review of a procedure or method. The closest some of the methods shown in Table 8.2 come to is an operational qualification, i.e., demonstrating that the task can be performed.

Table 8.2. Methods Used in Conducting a Technical Review

Method	Basic questions to be answered	Who is involved
Unison reading	• Does the current (or completed) version of the document match the earlier (or submitted) version?	• Documentation specialists

(Continued on next page.)

(Continued from previous page.)

Read-through	• Are the technical details of the document correct, current? • Are terms, equipment names, identification, etc. clear and correct?	• Content experts • Personnel experienced in the activity
Talk-through	• (Those above) • Is the sequence of actions correct, logical, and according to actual practice?	• Content experts • Personnel experienced in the activity
Simulation—experts	• (Those above) • Can the activity be performed as written by experts? • Is any critical information missing?	• Persons who are acknowledged experts in the activity • Persons who have hands-on experience with the activity
Simulation—novices	• Can the activity be performed as written by novices? • Is any critical information missing? • What supplemental skills, information needs to be provided by training, performance support systems, etc.?	• Persons without much experience in performing the activity • New/temporary personnel

Unison reading

This method is most useful when comparing two different documents that are *supposed* to be identical, such as a patient information leaflet that went to the typesetter and the "galley" document produced by the typesetter. The chief concern in this case is that no words or marks of punctuation are missing or lines duplicated. Unison reading is done in a quiet conference room allowing the two people to simultaneously read their documents aloud. The readers will, with practice, develop their own rhythm to their reading. If both documents are identical, nothing unusual will be noticed. If there is a word missing, there will be a very obvious, audible change in the pattern. This method is not useful in detecting defects in the technical content.

Read-through

The read-through can be done individually by content experts, with each expert focusing on areas applicable to his or her area of expertise (see Table 8.3). This approach is most useful for specifications, labeling, material safety data sheets, and other documents that are more information-based.

Talk-through

This approach is useful when the document describes the steps and substeps needed to accomplish a task, such as a standard operating procedure, emergency instructions, or a lab method. Content experts and those with experience in the activity would meet and review the document, with the leader walking the group through the process on paper. The experts should be able to identify steps that are out of sequence or where details are missing.

Simulation—experts

This method of reviewing technical content can be done using a simple "walk-through" with experts walking through the steps or it can be done while they are actually performing the task on "live" material. As in the talk-through approach, missing steps and technical accuracy are assessed. If "real" material is being used, there must be a clear understanding of what happens if there is a failure—the resulting product might, at the least, have a deviation that needs to be justified.

Simulation—novices

Sometimes procedure writers call this approach "the idiot test" or, more politely, "naive user test." For example, as wicked as it sounds, one group that writes instructions for using personal computers gives its documents to a senior management person (who is rather computer illiterate)—if Jerry can do it, anyone can.

In selecting novices, it is useful to have their basic knowledge and skill level be equivalent to new workers being brought in who would use the document.

Be very cautious in using novices if there is real product involved or if a failure has any potential of causing injury to people or damage to facilities or information resources. If a novice is attempting to access laboratory information, it may be safest to have a test database that is a copy of the production version.

According to a study, 34 percent of physicians have handwriting that is classified as fair to poor (*Journal of the American Medical Association*).

Table 8.3. Examples of Document Review Focus

Expertise	Assigned Areas in Document to Review
Operations	• Content detail • Sequence of activities • Names of equipment, materials • Rates, limits, times, etc.
Regulatory Affairs	• Agency requirements • What is currently registered with agency
Quality Unit	• What other sites/departments are doing • GMP requirements, issues
Validation	• Validation requirements • Validation status of current action

Results of the Review Phase

From the two stages of review, there will be a list of comments on where the draft document fails to meet its requirements. These problems can be categorized into 4 sets.

1. **Language, grammatical, spelling, and typographical errors.** Simple mistakes like noun/verb agreement, punctuation, a missing or misspelled word. If the typographical mistake involves a limit or the name of a chemical (e.g., *sulfite* instead of *sulfate*), the problem could also be categorized as a content error.

2. **Mechanical errors.** Not following the prescribed format or style, such as a heading that is incorrectly printed out as:

   Results of the review phase

instead of the proper way, as specified in the style guide:

Results of the Review Phase

3. **Content errors.** Technical problems with the information, such as missing steps, incorrect names for equipment, outdated information, improper sequence of substeps.
4. **Performance errors.** Problems appear while the document is being used that are, in part, because of deficiencies in the document. For example if a master manufacturing document has two different ways of recording weights (Figure 8.1), there will be confusion and mistakes.

Figure 8.1. Example of how inconsistencies in a document can lead to performance errors (Vesper 1993).

weight input lines on p. 3	weight input lines on p. 5
Gross weight: ______	Tare weight: ______
Tare weight: ______	Gross weight: ______
Net weight: ______	Net weight: ______

Routing the Draft Document for Review

Routing is the term that describes who receives the document and the sequence in which they receive it. The concept of routing applies to both document review and document approval activities.

Why routings are important

Understanding and then using routings effectively are important for several reasons. A well-planned routing can reduce the cycle time spent in document review and approval. A defined routing can make it easier to track missing documents as well as track the time individuals take to review it, something that can help monitor and improve the process. Finally, if the documentation activities will be moving to electronic management, distribution, and control of documents, routings will need to be developed to enable the computer system to control who sees the document and when they see it.

Types of routings

There are three types of routings that are commonly found in documentation systems: sequential, parallel, and hybrid. These can exist whether the system is paper- or electronically-based.

Sequential routing

Sequential routing, sometimes called "serial" routing is the simplest—the document is passed on from one person to the next in a defined order (Figure 8.2). Reviewers can list their comments all on one sheet.

While it seems straightforward, it is obvious that this routing is very dependent on each person's reviewing the document, commenting upon it, and passing it to the next person in the sequence. If someone is away and there is no one designated to review documents in his or her absence, the document will just sit there, holding up the entire process. Another problem with sequential routings is that the total time for the process is the sum of the time *all* the individuals take to receive/read/

Figure 8.2. A sequential routing.

comment upon/send out and any "queue" or wait time associated with it.

Parallel routing

Parallel routing sends the document to all of the reviewers at the same time; the originator then receives all of the individual comments (Figure 8.3). The time for a round of review is the time it takes the slowest individual to receive/read/comment upon/send out. Another advantage of parallel routings is that one can easily monitor the performance of each individual reviewer. While this method is easy to do electronically, it is more difficult to do in a paper-based environment.

A problem, also found to a lesser extent in sequential routings, is that a reviewer does not get to see the comments of the others reviewing the document. Collecting the comments and sending them to the reviewers allows everyone to see what was written.

Figure 8.3. A parallel routing.

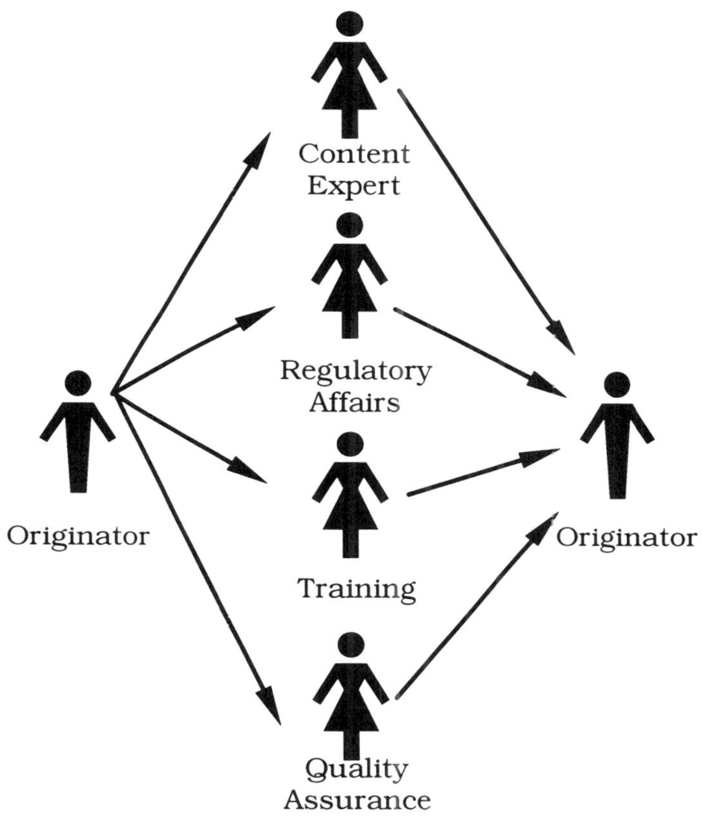

Hybrid routings (Figure 8.4) are a combination of the sequential and parallel approaches. These are useful when you know that certain reviewers will not need the information from each other as they make their evaluation. (They will see the comments eventually when the comments are passed back a second time.)

Figure 8.4 Example of a simple hybrid routing.

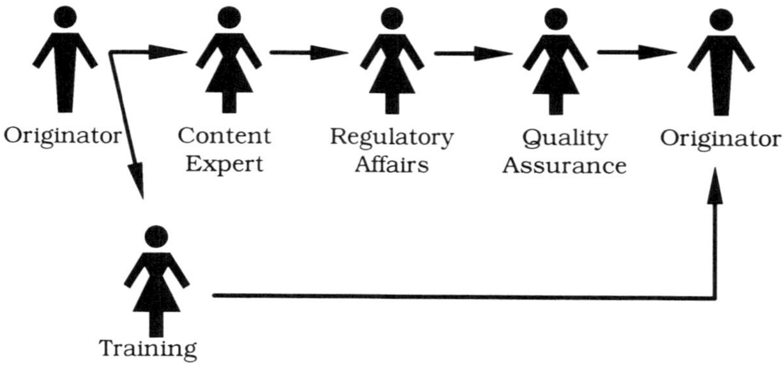

A variation on these three types of routings is a *subrouting* (Hollis and Meserve, 1994), in which a designated reviewer on the established route passes the document to someone who is not on the route list (Figure 8.5). This is useful when the designated reviewer wants to have others supply comments or if he or she doesn't have the time to perform a thorough review.

While subrouting might be done for valid reasons, it turns a formal, well-defined process into an informal one—control is lost. In sequential reviews, this is especially important as the length of time can increase.

Should subrouting be permitted? In most organizations, it is happening even if it is not formally acknowledged. There are significant benefits to the practice, but it should be controlled. Some ways of doing this include the following:

- Permit subrouting during the review process for specific reasons;

- Permit subrouting as long as it is conducted within the time allowed for the normal review (i.e., it should not add significantly to the overall planned review time);
- Identify people who have participated in the sub-route;
- Limit subrouting to the designated reviewer's organization, that is, a reviewer in QA should not forward the document for review in the materials management division; and
- Do not permit subrouting for the approval process (see the following chapter).

Documents Created During the Review Phase

There are different amounts of information that can be obtained and saved during the review phase. As men-

Figure 8.5. Example of a subrouting.

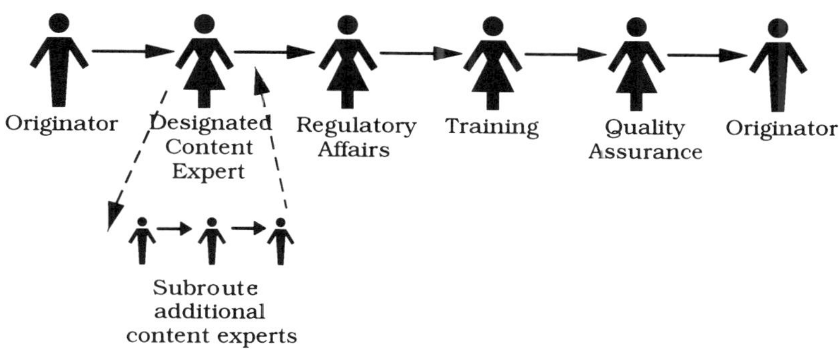

tioned in Chapter 2, the expectations of FDA inspectors vary considerably; hence, so does industry practice.

To start, consider what you need that will allow you to correct a document. For example, the comments collected will do the following:

- Identify and correct details, sequences that are incorrect from technical, quality, regulatory, etc. points of view;
- Identify elements that do not meet the format and style standards;
- Identify typos and missing information;
- Identify information that is difficult to understand;
- Serve as a catalyst for discussing issues that are controversial or haven't been thoroughly thought through;
- Communicate the findings to someone (the originator) who can make the corrections;
- Communicate the findings to others;
- Demonstrate that the document was reviewed by the designated reviewer; and
- Serve as a checklist upon which the changes will be confirmed.

The type of document used to record these comments will depend on a number of factors, such as if a sequential or parallel routing is used or if the documentation system is paper-based or electronic. Regardless, you will want to know:

- Where in the original document the change is to be made;
- What the new/correct information should be;
- Who wants the modification or correction;

- When the person reviewed the document; and
- If no modifications or corrections are requested, that the document is satisfactory to the reviewer.

In some situations, you may also want a justification for the modification.

The comments are then passed back to the originator for evaluation and, when deemed to be appropriate, modification to the draft document.

It is important to note that this review and the comments/discussion are concerned with how the information is communicated on the document, *not* whether a change should be made. The latter is part of change control and is discussed in more detail in Chapter 17.

Retaining comments

One of the questions about the documentation review process that causes heated discussion is, *Should comments be retained after they have been addressed in the preparation of a new draft and, if so, how and for how long?* Practice in industry is varied, particularly in firms using a paper-based system. A very small percentage of firms in the U.S. have been cited by FDA inspectors for not retaining the comments (see Chapter 2).

Why would you want to keep comments after they have been addressed? Some document coordinators say having them is very useful, particularly if a document is reviewed by many different people and/or is subject to churning (see "Churning," later in this chapter). Having comments available to show those who may be prone to make and unmake and then remake a change can be very useful. Comments are also useful if a particularly complex change is made and someone wants to know more details about why it was made and who was involved.

Another comment-related issue is, *Should the comments be saved after the document has been officially approved and is effective?* Again, industry practice is varied on this. Based on informal surveys at workshops, fewer firms keep them after approval than do firms who keep them while the document is being drafted/reviewed/redrafted. Firms who do keep the comments have said they are not frequently used, but when there is a need, they are very helpful.

What, then, are some of the options based on what drug and medical device firms are doing? Ranging from the least to the most stringent, options can include retaining the comments:

- Only for the draft cycle—they are discarded once they have been incorporated into the subsequent draft;
- For the entire review phase—once a draft is sent for approval, they are discarded;
- Through the approval phase—they are discarded after the document has been approved;
- For the life of that particular draft—they are discarded once a new version of the document has been put into service;
- For 1-7 years—discarded as part of a standard document destruction program; or
- Forever.

One other variation is to assign different retention periods to documents based on their importance to the organization. For example, for master production records or validation protocols, the retention time might be for the life of that document, while for procedures for sample handling, the comments are discarded after the document has been approved.

As with other GMP gray areas, it is important that you have a well-thought out rationale that documents what you are doing and why you are doing it. A policy would be an appropriate place for this information.

Controlling Documents During the Review Phase

There are two aspects of controlling documents while they are in review. First, they need to be individually identified so you can distinguish between the third draft and the fourth draft of a particular procedure, for example. This is discussed in detail in Chapter 5.

Second, the availability of the documents must be controlled, allowing only those involved in the review process to have access to them. This is important so that personnel are not faced with the choice of which document should be used. (A situation arose at a firm when an employee felt that "newer is better" and used drawings that were being reviewed. The devices made from the unapproved drawings had to be destroyed.)

Controlling access to draft documents

For paper-based documentation systems, one of the keys to controlling documents during the review phase is to limit those that are created and circulated: it is easier to control 5 copies than 15 or some unknown number. Simple statements on the top and bottom of drafts can help, such as:

DRAFT DOCUMENT. NOT APPROVED FOR USE.
DO NOT COPY.

Other controls firms use in distributing paper copies of drafts is to print them on green or pink pastel paper that, if photocopied, will show up as a gray. (Another advantage to this is that the colored draft copies are easy to spot on someone's desk.)

For electronically-based systems, draft/unapproved documents can be stored in separate files that require special passwords to access and/or print. Some documentation systems permit each document to have its accessibility limited to certain personnel. Having the DRAFT DOCUMENT statement, as above, on each page of electronic documents is also useful.

No matter what type of system is used or its sophistication, *somebody* will figure out its weak points and how it can be defeated. That is why all personnel need to understand why the controls are in place and the importance of not trying to thwart them. (Conversely, if people are defeating the controls, it would be instructive to learn why they are doing so . . . maybe they have a valid reason for using the draft documents but the overall system does not provide this functionality.)

Problems Seen During Document Review

No matter how elegant the system in place is, there are common problems with all documentation programs. Some of the more common ones are discussed below, along with some ideas for overcoming the difficulties.

Churning

Churning is the repetitious document cycle of draft, review, redraft, review, re-redraft, etc. There are some classic reasons for this.

Changes aren't included in the redraft. There is nothing more annoying for a writer to see revisions that have been agreed to or were noted, obvious problems not incorporated into the next draft. (This can fray the nerves of book writers and editors as well!) Marking each recommended change on the draft when it is made in a blue pen and then, when the change is confirmed, marking it with a red pen is one way of being sure that everything was addressed. If a change was not accepted, a few words to tell why are very useful; this should be shared with the other reviewers. One other variation of this is not using a "global" change. For instance a reviewer noting at the beginning of the document to change all instances of "Form 1572" to "Form FDA 1572" instead of marking each one.

Reviewers know there will be another chance. Some reviewers do a cursory job the first time they examine a draft, thinking that if there are any additional problems, either somebody else will catch it or they will have more time when it goes through the *next* round of reviews. One way to minimize this is to limit the changes that can occur beyond the second review. In other words, you want to make it difficult for people to make nonsubstantial changes after a certain point. (Some book publishers require that authors pay for changes they make after the pages are type-set.)

The process, organization, etc. isn't stable. One firm's quality assurance group was trying to define who was responsible for reviewing and approving deviations. Every time the draft procedure came back, job titles were changed or another position was added. The problem was that their organizational structure was volatile and it kept changing. This can also occur when a process is still new or process development personnel keep making "adjustments" to it. If a process is changing, try to write that part at a level of detail that is constant, not at a level of detail where changes are constantly occurring. When describing a position (e.g., the director of manufacturing), use the title, not the person's name.

The draft document is used inappropriately as the forum for a debate on the issue. The documentation coordinator from one company described how the technical services and the manufacturing group kept making comments on what the other was proposing. Each time they did this, a new draft had to be issued. Finally, the director of quality held a "summit" with all groups present. Only then, after the issues—which were serious—had been discussed and agreed upon, was there a consensus on what should be included in the document.

The appropriate reviewers were not defined at the start. Sometimes churning can occur because not enough thought was given to who are the most appropriate people to review and comment upon the document. This can happen when people subroute a document or make a note on the draft saying, "Has Kelly in Purchasing reviewed this?"

Not focusing on what is important. This is where everybody tries to be a writer (see "Wordsmithing," below) or go beyond their area of competence. Assigning reviewers things they are to primarily look for can help. Table 8.3 shows an example of this. Training sessions for reviewers can also help.

Formats and styles are not agreed to at the start. These types of changes are very annoying—for example, when a reviewer wants a heading to be in italics instead of roman type, or when someone wants to use "Attachment A" instead of "Attachment 1." Having a detailed style guide can reduce some of these comments.

Also, a glossary of "official" terms and abbreviations is useful.

"Wordsmithing." Wordsmithing occurs when a reviewer insists that "use" be replaced by "utilize" or "to describe how QA will . . ." be changed to "to describe the procedure by which QA will . . ." A well-trained editor or technical writer who wants documents written understandably should have the final say on this issues.

Documents lost on a reviewer's desk

Some reviewers work in a chaotic environment, their motto being, "A clean desk is the sign of a sick mind." While some of these people can instantly put their fingers on any document, others aren't quite as talented; documents in the review or approval stages vanish among all the other paper. One firm solved this by attaching a garish neon-colored cover sheet to all of their documents, making them almost impossible to misplace. A side benefit: people reviewed them quickly to get them off of their desks.

Review periods taking too long

Most firms who are using paper-based systems do not track how long it takes for each review/redraft cycle, although most documentation coordinators can tell you that it takes longer than it should.

Before criticizing the reviewers, a documentation coordinator should be sure the system is properly designed. Are the routings optimized? If a person isn't at work, is there a designated substitute for him or her? Does the substitute know what he or she is to do?

One of the simplest ways is to establish and enforce a turn-around time: establish a time reviewers have to review and return their comments. If the document hasn't been returned, the reviewer is notified, with a copy of the message sent to their management. Yes, it is a hammer approach, but it works.

In some situations, document review is viewed as low priority item. That means that management feels it can live with outdated information. Sometimes they need to know what problems this is causing.

One other way of addressing the review time issue is to completely change the approach. Some firms have *review parties* where, for example, reviewers receive copies of the documents on Monday afternoon and need to read them by Thursday. Then, on Thursday afternoon, all reviewers attend a standing meeting where the documents are discussed with the goal of making them "approvable." If a document needs significant additional work, a subteam works on having it ready for the following meeting. This approach does take commitment on the part of the reviewers and management, but it can significantly reduce the cycle time. If this method is used, someone should keep minutes of the decisions and assignments; the minutes should be retained.

Conclusion

At the end of the review process, you should have a document that meets the standards established for its format and is technically and grammatically correct. In other words, *it is fit for use*. A reviewed document *may* not be perfect. The quest for perfection is often why it takes writers, reviewers, and designers so many iterations.

A new document reviewer in a regulatory affairs group at major drug manufacturing firm became reluctant to sign her name to anything because she was fearful there might be something wrong with the document that she didn't catch. Her supervisor patiently asked if she had done her work according to the procedures, if she had participated in the recent training activities, if anything she saw indicated there were problems, and if she had documented her work. Her answers, said aloud, convinced her that she had done things properly and, to the best of her knowledge, when she signed the document, everything was correct.

The same holds true with reviewers of functional documents. They need to work for excellence, not for perfection.

Frequently Asked Questions

Q. *Our plant site uses review meetings to discuss and finalize documents, but we still have churning—we can't get people to agree. We're getting desperate. What can we do?*

A. Desperate times require creative solutions. One way is to force as many of the changes to occur in the first review round as possible. Having a "penalty" for subsequent changes can make the point. One firm gives its reviewers three Ping-Pong balls that they use during a weekly review session. For documents having their first review, the reviewers can make as many changes as they want. For documents on their second review, reviewers can make changes if they want, *but* they have to give up a Ping-Pong ball. When they are out of balls, they are not permitted to make additional changes, unless they can talk colleagues out of one of theirs. Quarterly, those with the most balls get some simple gift like a funny T-shirt. This has added some fun to the process, but more importantly, it has improved the effectiveness and efficiency of the reviews.

Q. *Our review process handles documents by exceptions, that is, if comments are not returned, we say that the reviewer has no problems with the document. Is this OK?*

A. No. Firms have been cited for this practice (see Chapter 2). Even if someone has no problems with a document, this needs to be communicated to the documentation coordinator. If people are not returning review documents because of other priorities, lack of interest, etc., management needs to help enlighten them to the criticality of their role in documentation.

Q. *How many people should be involved in review? We're thinking the more the better. Is this correct?*

A. Bigger isn't better when it comes to the numbers of people involved in document review. Keep it simple. Think of the value each person will add to the process. If large numbers are needed, consider having a discussion of the document off-line, that is, apart from the review process. Also, remember the requirements/expectations of the medical device Quality System Regulation (and ISO-9001): "Changes to documents shall be reviewed and approved by an individual(s) in the same function or organization that performed the original review and approval, unless specifically designated otherwise . . ." (21 CFR § 820.40 (B)).

CHAPTER 9

Document Approval

Goals

- Define *document approval*.
- Differentiate between review and approval.
- Identify who is involved in document approval and what they do.
- Discuss using different levels of approvers.
- Discuss how routings are used in document approval.
- Describe different options to provide document approval.
- Discuss dates related to approval and what they mean.
- Discuss control issues related to document approval.

As you enter the approval phase, you will have a final draft of a functional document that has been reviewed, commented upon, and is, in the minds of the reviewers, correct in the following areas:

1. Language, grammar, and spelling;

2. Mechanics, such as format and style;
3. Technical content and sequencing of steps;
4. Performance support, i.e., that this document contributes to the accomplishment of the task, decision, etc.

What you do not yet have is an official document that says the department, laboratory, or company will do or make decisions according to what is contained in the functional document. This, then, is what happens in document approval: Designated organizational personnel authorize the implementation and use of the particular functional document within the document's scope and timeframe. Approval also means that management will take the actions needed to communicate the contents to those directly and indirectly affected by it and provide support (training, coaching, etc.) as needed to make it happen. As *Star Trek's* Commander Piccard says, "Make it so."

Differences Between Review and Approval

Some organizations treat document review and approval as one step. While some of the reviewers/approvers may be the same in such situations, the thought processes are different. Review looks at the technical side of a specification; approval considers more of the broader organizational issues. Table 9.1 illustrates this.

Table 9.1. Comparison of Review and Approval Phases

Review	Approval
• Content is correct. • Information is provided in a logical, proper sequence. • When done according to this description, the task will be accomplished/the product will meet its requirements. • From a technical point of view, this is what should be done.	• Confirms that the review process occurred and was adequate: the document is, to the best of the approver's knowledge, correct. • From an organizational standpoint, this is what we will do/what we expect. • Efforts such as vendor audits, training, and coaching will be used to accomplish this. • Anything contrary to this will be considered a deviation and be documented and investigated. • Those affected by this document will be informed, trained, and provided with any other required tools by the effective date.

Specific Expectations of the Approver

As the approver examines the document, he or she needs to know the basis on which it can be approved. If they have content expertise, it is assumed they are approving it based on their own knowledge of the correctness of the content. However, if the document is being signed by a middle manager, he or she will need to rely on the review conducted by technical experts. In this case, the reviewer acknowledges that the review occurred. Table 9.2 provides examples of the basis for approving a document.

Table 9.2. Examples of Bases for Approving a Functional Document

Who	Basis for approval
Manager of area that will be using the document	• That content is correct, either by direct knowledge or that the appropriate expert(s) have reviewed and commented upon the document. • That the document will be used. • That all preliminary actions (training, communication with vendors, etc.) will be completed before the effective date.
Regulatory Affairs	• That the document is consistent with the agency-approved submissions or that the appropriate notifications will be made.
Development	• That the specifications, requirements are correct and based on data, good science.
Quality Unit (QA/QC)	• That the document meets current GMP and agency expectations. • That content is correct, either by direct knowledge or that the appropriate expert(s) have reviewed and commented upon the document. • That the effective date is reasonable and appropriate. • That the document will be used.

Document Approvers

The documentation procedure needs to define who is authorized to approve documents. There are several considerations in this. The approver must have at least *one* of the following:

- Responsibility for the organizational area and/or the domain of the issue,
- Authority to make the decisions for that area/domain, or
- Authority to provide resources that will support the implementation and continued use of the document.

Responsibility for the organizational area and/or the domain of the issue. The organization or regulatory agency would hold the person *accountable* for the area, product, or domain. For example, one of the approvers for a specification for plastic bottles should be a manager in production, where the bottles are used.

Authority to make the decisions for the area/domain. The approver has been given, by the organization, the right and power to permit the document to be used and enforce its use. Using the specification example again, the approver has an organizational connection to the acquisition and use of the new bottles.

Authority to provide resources that will support the implementation and continued use of what is required in the functional document. For example, if a procedure is written that requires GMP training to be given to all personnel on a quarterly basis, those approving the document will provide the resources or reorder the priorities so the training will happen.

Effective Dates and Approval

The *effective date* is when the document is "turned on"—for example, when the specification will be used replacing the earlier specification or the first time an SOP will be used. During the approval phase, management from different parts of the organization need to confirm that the proposed effective date is appropriate and reasonable. Approval date does not necessarily mean effective date. Typically, there are things that must happen before the document can go into effect. Figure 9.1 shows this graphically.

Appropriate means that the effective date is consistent with decisions made in the change control process and/or that the date the document is first used is consistent with regulatory requirements such as preapproval or simple agency notification.

Reasonable means that there is time to complete all the preliminary activities, such as distributing the document, informing or training personnel affected by the document, and having any other necessary documents or tools available for use. In other words, everything must be in place so the document can be safely, efficiently, and effectively put to use.

When management signs the document approving it, they are also approving the point in time when the document can first be used. Management should be aware of and address any issues that may complicate its use on the designated effective date.

Figure 9.1. Documentation timeline.

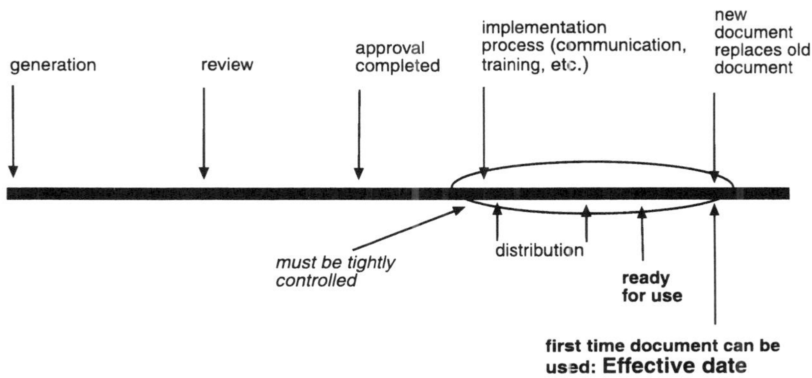

Start Point and Approval

Medical device firms have another point at which they sometimes designate a document takes affect. This is sometimes called the *start point*. This could be when the previous items are exhausted from inventory and the new version is ready to be used. It is critical that there be tracability of what documents were used, along with the parts, items, process, etc. used to produce a product.

The Role of the Quality Unit

One of the underlying concepts of good manufacturing practice is that the quality control unit (QA/QC) has the responsibility of approving all types of functional documents (see Chapter 2). QA/QC needs to be sure all quality and GMP issues have been addressed and satisfactorily answered. While other approvers may change based on the type of functional document and its contents, there must always be an approval given by someone speaking for the quality unit.

GMP auditors and agency inspectors during audits will test for this by asking, "if there is a disagreement, who has the final say?" They will want the final word to be made by the quality unit.

The role of QA in the medical device industry is slightly different; QA's role is important, but the new Quality System Regulations allows QA's role to be more diffused throughout the firm. Also, the firm's Quality Manual defines how QA actually operates in the firm. Because of this, QA may or may not be required to sign off on all specifications, procedures, methods, etc.

How approval differs from change control

Differentiating between approval to make a change and the approval of a functional document can sometimes be confusing. Simply put, the document (specification, master production record, SOP) is only the symbolic description of the change, not the change itself. For example, if a primary container is being changed from one type of plastic to another, a change committee would be involved in evaluating the proposal, the benefits/risks, the need for any stability data, etc. Their decision—to go with the new plastic—is a change in the *content* of the related specifications, packaging instructions, etc. (Chapter 17 discusses changes in more detail.)

Approval Hierarchies

During the review phase, content experts evaluate, among other things, the technical details presented in the document. If a new cleaning procedure for a particular piece of equipment is being adopted to cover multiple departments, for example, it is a good idea to have various stakeholders review the SOP to be sure that all areas can accomplish the task as described.

From an approval point of view, there are a couple of different approaches that can commit departments to what is described in the particular document. One way would be to have all of the managers for the individual departments sign as approvers, saying that their areas will comply with the new cleaning SOP. Another approach would be for the manager's manager, e.g., a general manager, to sign, indicating that all areas in his or her realm of responsibility will meet the requirements.

In general, the more encompassing the functional document, either the more authority/responsibility required of the signer or the greater the numbers of signers required. Table 10.3 shows an example of an approval hierarchy.

The status of the approver also signifies the importance of the document. A quality statement that affects the entire company carries more weight when signed by the chief executive officer than by the department head of QA/QC. (One important caveat: if the person is not recognized by the members of the organization to have a commitment to what he or she signs, the result will be confusion or cynicism.)

Come Back in Two Years: I'm On Hold

Michael Fortino, a time-management expert, concluded from his firm's studies that the average American will, in a lifetime, spend five years waiting in line, one year searching for belongings at home or office, three years attending meetings, and eight months opening junk mail (*USA Today* (6/24/88)).

Table 9.3. Example of an Organizational Hierarchies and Document Approval Chart

If the document affects this organizational area,	It should be approved by management at this level:
Entire organization	• Executive vice-presidents for Technical Operations, R&D, Marketing, Administration, Human Resources OR • Chief Executive Officer
World wide manufacturing, packaging areas	• Executive vice-president for Technical Operations OR • Executive director Technical Operations • Director, Product/Process Development (as needed)
R&D	• VP or Executive Director
Local sites	• General manager OR • Directors of affected divisions
Division	• Director
Departments	• Manager

Using Routings for Approval

In the previous chapter we defined a routing as the designated persons who receive a document and the sequence in which they receive it. This definition applies to approval process as well as review. Approvals generally use a sequential method with operational areas signing first. The final signature should be someone with the appropriate organizational level in the quality unit.

Conclusion

Once the document is approved, implementation for its use can begin. Approval is more than just stating that the content is correct. Approval means that the organization will use it and that any deviations from it or that occur while using it will be investigated and documented.

Frequently Asked Questions

Q. *Our company's Director of Quality is too busy to review and sign all of the SOPs and specifications. What are the alternatives?*

A. The procedure describing document approval should include a provision for an alternate approver. In this situation, the Director of Quality would designate a subordinate to read through the document and sign. You might also want to use this for other management people if they are on vacations or traveling.

Q. *During the approval phase, some of our management want to "fine-tune" the documents, forcing them back into the draft phase. What can we do?*

A. If an adequate review is done initially, the documents that the approvers get should be correct and written according to company standards. If this review is not occurring properly, and there are mistakes, the management is right to send back to be revised. On the other hand, some managers are frustrated editors and want to make their mark on a document. If this is happening, try one of the following suggestions:

1. Make sure they understand who reviewed the earlier drafts and that the final draft was deemed to be correct.
2. Let them know how this is delaying the process and that any changes really need to add value to the end document.
3. If they insist on making changes, include them in the review phase.

Q. *Can the same people approve a document as those who initially reviewed it?*

A. Yes. There is no problem with this as long as they are competent in the area they are reviewing and they have the organizational authority and responsibility.

Chapter 10

Document Distribution

Goals

- Define *document distribution*.
- Differentiate between *push* and *pull* systems.
- Identify who is involved in document distribution and what they do.
- Discuss who uses documents and when they need them.
- Distinguish between master documents and controlled copies and reference copies.
- Identify and discuss methods for document distribution.
- Discuss issues related to document distribution.

Document distribution is the process of providing the newly approved document to people who will be using it. Distribution must be done in a way so that:

- The document is readily available at the point of use to those using it,
- There is no confusion over which documents are or are not to be used, and

- There is a way to reconstruct what document was used in a particular task or decision.

The Distribution Process

Once the document has been approved for use, the process of making it available to those needing it can begin. A number of firms have been cited by regulatory agencies for not having the proper controls in place for distribution. For example, one firm had multiple versions of the same laboratory method available to the analysts. Another firm had provided its engineering group with procedures that were rapidly changing; there was no process for keeping the engineers abreast of the changes.

A challenge in distributing documents is that there are multiple users and user needs that must be satisfied; some of these needs can potentially be in conflict with each other. For example, lab personnel need to be trained on the new methods but cannot use them for actual product-release testing until the effective date. In device manufacturing, the new assembly instructions may be valid only for making devices for one customer while another customer has its devices built using the older (but still valid) set of instructions. The distribution process needs to address the needs of the users and at the same time, prevent confusion among those using them.

Two Methods—Push and Pull

There are two basic ways of providing people documents, using a *push* system or a *pull* system.

Push system. This is the more common method that is used in distributing documents, particularly with paper-based documentation programs. Here, the document coordinator provides copies to those who have been identified as primary users and implementation personnel for that document (see below). The mechanics of the push system are discussed later in this chapter.

Pull system. As electronic systems become more prevalent, this method of distribution is being used more frequently. Here, the primary user asks for a copy of the document. This can be done through an input screen on a computer network or selecting a document from a read-only "library." Firms with paper-based systems usually have pull systems for one-time and reference users: when these users want a document they call and have one issued. When electronic document management systems are used, permissions or security levels need to be set as to who in the organization has access to the document.

Advantages/disadvantages to push and to pull systems

Push systems are easiest when there are a relatively small number of users or points of use and when the documents are targeted to known users. (One firm had a "shotgun" push system but they did not target their procedures, meaning that *every* department head and manager received a copy, whether or not it applied to his or her area or not.)

A disadvantage of a push system is that potential users could be missed, especially if the organization is a large one or the documents haven't been targeted precisely. (The shotgun approach mentioned above is one way to *not* miss anyone, but users may become saturated with documents that don't apply to them. Also, a lot of trees die needlessly.)

The pull system works best with one-time and some reference users. People know that if they need a document, they contact the document coordinator. (Approaches to controlling issuing documents under these circumstances are discussed later in this chapter.)

A pull system also works well if documents are available electronically. If viewing the document on a computer screen is adequate, users can sign-on and get the most current version whenever and wherever they want. For example, one firm's QA batch release personnel were required to retrieve an electronic copy of the current batch release SOP every time they released a batch of product.

Other options are available if a paper copy is needed. Another firm, which had a standard effective date of each Monday for its new SOPs, had its supervisors download all the SOPs their people would use that week. Personnel made use of these paper copies, which were then destroyed at the end of week. The following Monday, the supervisors would again retrieve the most current set.

Having access to electronic copies is one of the best ways of assuring that people have the most current document available. This way, there are no questions: the system is the library for current information.

People Involved in Distribution

Most firms have documentation coordinators who are assigned the responsibility of providing documents to those needing them. Coordinators use either a paper-based, computer-based, or some sort of hybrid system to assist in the process. These systems need to be carefully thought out and strictly adhered to, otherwise there will be chaos. There is not one system that can be copied and applied to each firm; every firm is different and has specific needs. There are, however, things that can be done to create a workable system for your organization; these are presented in the steps that follow.

There are two parts to the process of defining the users: 1) Who and 2) When. The first step involves identifying *which individuals should be receiving what* category of documents. If a person gets a particular category of documents, the second step considers *when they receive the document*.

As you identify the users, you will need to determine if they need to have their own copy of the document or if several users can share the same copy. This is particularly important if a paper-based system is used.

Identifying the document users

The originator of the document should be the initial person to identify the users of the document; this usually relates to why it was created and who will be using it, either directly or indirectly. In other words, this matches the document to the user. A user matrix can be developed to assist with this (Figure 10.1).

Typewriter, 1867; United States

Mark Twain was the first author to submit a typed manuscript, which he wrote on a Remington soon after the machines went on sale, for a costly (at the time) $125. The shift-key typewriter, which introduced lowercase letters, appeared in 1878, and "visible writing," which let the writer see the printed line as it was typed (instead of having to lift the carriage to check a letter or word), originated the same year (Panati, 1984).

Figure 10.1. Example of a document-user matrix for procedures.

Keyword	Dry Prod Mfg.	Liquid Prod Mfg.	Packing Labeling	QC Labs	QC Batch Release
Analytical method				X	
Equipment cleaning—general	X	X	X	X	X
Equipment cleaning—specific types	Defined by originator	Defined by originator	Defined by originator	Defined by originator	Defined by originator
Preventative mainte-nance—general	X	X	X	X	
Preventative mainte-nance—specific types	Defined by originator	Defined by originator	Defined by originator	Defined by originator	Defined by originator
Stability program—sampling	X	X			X
Stability program—testing re-quirements				X	

Defining when people need the document

Once you know which documents are applicable to users, the next step is determining when they need it. Users can be categorized into four different types:

Primary users: Persons who will directly use the document as they make a decision or take an action. Examples of primary users are purchasing agents, requiring specifications as they place an order with a vendor, and a component assembler who uses a drawing in building a medical device.

Primary users need to have available the documents that are effective *at the time and place* ("point of use") they are doing a task or making a decision.

Implementation personnel: Persons who need to use the soon-to-be-effective document as they prepare for when it becomes effective. This group may include lab personnel who need to order new reagents or growth media that are required by a revised laboratory method. Another example is the departmental trainer who needs an advance copy to begin to create training courses on a newly validated cleaning procedure.

Implementation personnel need to have the upcoming versions of the documents available to them so they can adequately prepare for the document's eventual usage.

One-time users: Personnel who need the current version for a one-time, single use such as an internal audit procedure that is given to a subsidiary for their information or a cleaning procedure that is provided to an engineer who is purchasing new manufacturing equipment. One-time users typically have a particular question they need to answer.

One-time users need only the version that is in effect when they request the document. They do not need to have it updated.

Reference users: Personnel who need the document to answer a question, for example, a regulatory affairs specialist who needs to know the past and current specifications for a packaging material as they prepare an annual report to a regulatory agency. In most cases reference users will use controlled copies.

Reference users need to have rapid access to the current and sometimes historical information on a potentially wide variety of topics.

Many of the same issues related to the pushing and pulling of documents come into play when considering when people can access documents. If a push system is used, the document needs to be pushed to them at the appropriate time (neither too early nor too late); if a pull system is used, provisions need to be made so people do not have access to the document before they are authorized to do so.

A Quick Summary . . .

Just to summarize the major points so far:

- Documents can be provided to the users in either a push or pull approach. An organization could use either one or both of these in providing documents to the users.
- The initial considerations in distributing a document are determining who gets the document, if they need their own personal copy of it, and when they should receive or be able to access it. Document users can range from those who would use the document repeatedly as they perform a task (primary users) to those who would use the document infrequently or only once.

Master, Controlled, and Reference-Only Copies

Whether documents are produced using a paper or electronic system, there are usually three ways that documents exist. The first way is as the *master*. The master is the parent document; all other versions come from it. It could be in an electronic version from which copies are printed or in a paper version (or "mechanical") that is kept in a file cabinet and is the source from which photocopies (or blueprints) are made.

From the master, either controlled or reference copies can be produced. In a sense, all copies are controlled, but true *controlled copies* are replaced when new versions of the document are issued. The master is considered the *copy of record* for archival purposes, unless another is formally designated.

Reference-only copies, on the other hand, are not replaced. If the process, specifications, or part numbers have changed since the reference copy was issued, there is no way of knowing this. In other words, the content in a controlled document is dynamic; in a reference document, it is static.

A document can also become a static, nonchanging one if a copy is made of a controlled copy and no one is informed. For instance, a visitor from one facility visiting another site asks the department head for a copy of a cleaning procedure. The department head obliges, making a photocopy of her SOP for the visitor. An uncontrolled copy has been produced, which may be fine if it is meant to serve as a reference copy, but if it is put into service at the visitor's plant site, the document coordinator does not know that they should receive updates to the SOP as they become available.

One of the control features of your distribution system is for people to know if they have a document that is controlled, i.e., that it will be updated if a new version becomes available. For paper-based systems, this is done in a variety of ways.

Official copy stamp. A rubber stamp (usually made using red or blue ink) that states that it is an official copy issued by document control (Figure 10.2). This is useful when there are only a small number of copies issued. If someone makes a copy of this document, the stamped message is in black or gray.

Official paper. Copies printed on special paper, such as colored stock or paper that has a colored company logo or stripe. One firm uses paper that has a light blue stripe on the left side that if copied will not be visible. This special paper would be available only to personnel who can print the official documents.

Watermark. Most word-processing packages have the ability to create watermarks or light-gray printing on documents indicating it is an original. The watermark, which would appear on every page, could simply say OFFICIAL DOCUMENT or be the company logo or graphic. For this to be an effective control, it needs to be printed in such a way that it will be visible on the original, but not apparent on any copies. Experiment with the watermark's color, line pattern, and the printers that will be used.

Figure 10.2. Example of an "official" stamp placed on a document.

CONTROLLED DOCUMENT

Subject to revision

Omega Healthcare Products QA Department

*If this mark is not **red**, this is an unauthorized copy.*

Distribution Methods—Paper-Based

How do you get a document into the hands of the users or to the document stations (a centralized point where people go to access hard copy documents)? There is a variety of ways, depending on whether a push or pull approach is used and on who the user is. Distribution methods discussed below are ranked from the least controlled to the most controlled.

As you establish a distribution system you will be facing a paradox. You want the distribution method to be simple yet controllable. Also, you want the proper distribution channels to be easy for everyone to use so as to discourage people from taking shortcuts. For example, if a person needs a reference copy of a document, you want to make it fast and convenient for them to get an "official" copy, not to go off and create an uncontrolled one.

Push system

As mentioned earlier, this has been the most common approach for distributing documents to primary users, implementation personnel, and reference users.

Uncontrolled distribution. This method would send out a hard copy of a document to all identified primary users and require them to file it correctly, update their binders, etc. There is usually a generic cover memo that is sent with the document. The best thing that can be said about this method is that it is easy to do.

On the negative side, this method does not provide any immediate feedback to the documentation coordinator to assure that the document was received, that it was correctly included in the document collection, and that, if the document was a revision, the previous version was removed from service. This could also be called the "spray and pray" approach to document distribution. If the document is important from a business or GMP perspective, uncontrolled distribution should not be used.

Return-receipt distribution. This method is similar to the method above, but it requires the recipient to sign a receipt saying that the document was received and properly filed. The receipt could be part of the cover memo for the document (see Figure 10.3) or a specially designed form on colored paper. Another option is for the receipt to be e-mailed to the documentation coordinator.

There are several variations to this. If the document is a revision, it may be useful for the recipient to certify that the previous edition has been destroyed or, for more control, that the previous edition is attached to the receipt.

As the receipts are returned, it would be the document coordinator's responsibility to track the receipts to be sure that (1) the user received the document and (2) if it was a revision, the previous version was destroyed or returned. This can be done using a simple paper log or in an electronic spreadsheet or database. Whatever method is used, it needs to be kept simple and easily maintained.

One other variation on this method is to use bar-codes. A bar-code would be added to the top of the first page of the document using a mail-merge program; each copy is "personalized" to the user. When a revision is issued and received by the user, he or she returns the previous edition to the documentation coordinator, who would scan the bar-code into a PC. This would update a database, showing that the user has returned the previous version.

The return-receipt approach has more control than simply sending out documents, particularly if previous versions are returned. On the negative side, there could be a large amount of paper (i.e., the receipts and previous editions) going back and forth in an organization. If the previous editions are not returned, there is no assurance that the new copy has been placed at the designated documentation stations; the recipient might just put them in a pile to file when he or she gets the time. Another drawback to this approach is that there can be a lag time between when the documents are distributed and when the receipt is returned. In large organizations this could be several weeks. Also, there can be a delay between when the documentation coordinator receives the receipt and when he or she logs it in.

This approach can be effectively used in a GMP setting if it is closely monitored; there are several ways to do this. First would be monitoring the receipt log. Are there any holes in the log that would indicate that the document was not received or that the previous document was not destroyed and/or returned? Also, look at the dates of the log entries. If there are long gaps between when the document was first issued and when the receipt was received, it might mean a lack of control.

The other monitoring point is at the point of use. Do users have the proper documents available? Are they using the version that is effective at that time?

Courier delivered. Courier delivery requires more personnel, but it is one of the best ways of assuring that the correct document is in the right place at the proper time. Here, a documentation coordinator or clerk personally distributes the new documents and collects any outdated versions. A log is kept noting that this exchange was made. For this process to work, each documentation station must be known and accessible to the documentation coordinator.

One smaller firm that uses this approach saves all its distribution until late Friday afternoon. The two documentation clerks have a route they follow giving out the new and collecting old versions. The new documents all have an effective date of the following Monday.

Larger organizations might have a variation on this approach. Different divisions might each have a documentation assistant assigned who would get a bundle of the documents and then act as the local courier. In this case, the fewer couriers involved, the greater the control of the process.

"Prepublication" copies. This approach is used for supplying implementation personnel with the documents they need in preparing for when the document becomes effective. Since there are fewer people receiving these copies and because this is not an actual point-of-use, controls can be less stringent. It may be enough for these copies to be on special paper or stamped as "information only." Monitoring and auditing can determine if additional controls need to be established.

Figure 10.3. Example of a return receipt for an SOP.

RECEIPT NOTICE

SOP 34-532.1

Pest Control for Bulk Manufacturing Facilities

To: Document Control Coordinator
Mail Stop 31/A

The above named procedure has been:

1. Received and
2. Placed in the official SOP binder.

The former version has been securely destroyed.

Your signature means this has been satisfactorily done.

(Your signature) ____________ (date) _________

Pull system

By its nature, a paper-based pull system is easy to control because those who want the documents ask for them. Typically, these are people requesting the documents for one-time or reference use. Your distribution system should be able to accommodate these requests quickly and securely. If a person is told to wait, for what to them is an unreasonable amount of time, he or she will find another source for the document.

Defeating the System

In workshops about documentation systems, document coordinators usually talk about ways that people could bypass their controls. "What about color copiers?" or "What if other people can 'forge' a watermark?"

Yes, these are all possible situations, but if you make it easy for people do to the right thing, most people will. Part of the introductory and reinforcement training provided to document users needs to explain why these controls are in place and how the controls actually *help* them do their work with more confidence. Audits and spot monitoring are also useful in determining if people are working around the controls. If it is discovered that people are bypassing the system, the reasons need to be investigated with corrective (and disciplinary) actions taken as appropriate.

Distribution Methods—Electronically-Based

An irony of having documents available to users electronically is that it makes distribution easier *and* much more complex. People can go to the document files and get the most current version, but if hardcopies are or can be made, the system can quickly go out of control.

The key question to ask as you design a distribution method for an electronically-based system is, *where will users find the official copy*? Will it be on-line or will it be printed and available in a hardcopy form or will it be some combination of the two? Some user areas, for example, manufacturing operations, may not have computer terminals conveniently available at all points-of-use; therefore, paper copies may need to be printed and controlled. Other people, such as regulatory affairs professionals, may be able to get any document by way of the computer network.

One way of keeping control would be to limit the ability to print hard copies of the document. Most electronic document management systems have this capability or a simple program can be written allowing only those with passwords to print out the document. (There are applications and steps that can bypass this: "screen images" are one way of doing so.)

Another option would be to designate specific printers loaded with official paper as the output devices. Watermarks also could be used or headers on each page that give a message like "Do not use after (date)" or "Reference Copy Only."

Conclusion

Distributing documents puts them into the hands of the users. The critical part to all of this is that, at the point of use, users do not have to decide which version is the right one. You want to make it simple for them. Having a well defined, organized, and monitored program will ensure that people have the informational tools they need to do their jobs safely, effectively, and efficiently.

CHAPTER 11

Retention and Destruction of Documents

Goals

- Define *document retention*.
- Define *document destruction*.
- Discuss requirements for retention.
- Describe methods of retaining documents.
- Identify topics that should be considered in a document management procedure.

After documents are generated and used, after records and reports are prepared, what happens to them? What is kept? For how long? In what form? What is destroyed? When and how?

These are the questions that are addressed by *records management*, an activity (and profession) that establishes and implements procedures on document retention and destruction. In this chapter, we will build on some of the principles of records management but sharpen the focus to how these principles apply in healthcare manufacturing.

The Document/Record Paradox

A dilemma facing documentation professionals is that, while it can be useful having a great deal of information available, it is costly: there are direct and indirect costs associated with maintaining documents and records. There are also legal concerns with having documents that have been saved, by plan or by chance, for years and years and then found during the legal process of *discovery*, when potential evidence for a legal case is systematically searched for.

On the other hand, documentation is needed for some very important business and quality purposes, many that extend far beyond GMP compliance; you can have business and legal problems if information is not available. As mentioned in previous chapters, functional documents describe how something is (or was) to be done (e.g., an SOP) or the standard of what is acceptable (e.g., specification). Records present what actually occurred and who was involved; reports can include a summary of various records, the rationale for a decision, and what action needs to be taken. Without documentation, records, and reports, there is no evidence to show that GMPs were followed in the manufacturing, testing, and release of a product. If the records do not exist, in the eyes of most regulatory agencies and quality auditors, what was claimed didn't happen.

In this chapter, unless specified differently, the term *document* will include functional documents, work instructions, records, and reports, as well as drawings, photographs, e-mail, images, etc. Also, a document is any *form* of the information, whether on paper, on tape, or in a digital (computer) representation.

The Basics of Document Management

Stripped to its fundamentals, document (records) management is concerned with these essential points:

- Knowing what documents and records an organization has and where they are kept;
- Knowing what records a firm needs and how long they are needed for business, legal, financial, and historical purposes;
- Saving only the copies needed using the most appropriate, cost-effective methods and technology so they can be accessed when needed;
- Destroying what is not needed in a systematic way; and
- Creating and managing a system for routinely purging files of outdated, unnecessary documents.

In addition, most practitioners in the field also are concerned with control of data input forms as well disaster recovery.

As you see, records management extends far beyond GMP documentation; it is used to protect and manage a firm's information, easily one of its most valuable assets. Several sources for more information are books written by Gloria Gold, Susan Diamond, and Donald Skupsky; these three works are described in the References and Readings section.

Terminology

Some terms used in records management that are important as we discuss retention/destruction of GMP-related documents are the following:

Active record: a document that is used or is being frequently accessed.

Central file: a central location where documents are deposited and available for use/review.

Copy of record: the "official" version of the document that is subject to the document retention policy. (This presumes that other copies that have been generated have been properly destroyed after use.) The copy of record may not be the "original"; for example, the firm's copy of record of a letter sent to the FDA may be a copy retained by the person (or department) who wrote it.

Custodian: the person (or title) in the organization responsible for holding a particular document.

Decentralized file: a location, close to where documents are generated or used, that contains the document.

Document center: the location where inactive documents are stored and maintained.

Document inventory: a methodical audit, resulting in a detailed report, of what documents are kept by whom and in what form of media.

Document retention schedule: a detailed plan for where and how long particular documents are to be kept before being destroyed.

Document series: a group of a particular type of document (e.g., SOPs, specifications) that are stored together.

Disposal date: the point in time when the document is no longer needed and can be destroyed. The disposal date would be included in the document retention schedule.

Inactive record: a document that is not being used but must be retained for regulatory, business, or other archival purposes.

Purge: eliminating those documents that no longer have business, regulatory, or archival value to the organization.

Vital documents: those documents that are essential to an organization's survival. Would typically include certain financial records (e.g., contracts, accounts payable), research notebooks, and product-specific documents.

With this background, let's how see how we create and implement an effective plan for the retention/destruction of GMP-related documents.

Legal Considerations Related to Document Management

There are enormous legal and regulatory issues connected to document retention and destruction; Skupsky (1991) covers the topic in a very accessible yet thorough way. (His *Recordkeeping Requirements* is an essential book that every documentation group should have and read.)

Some particular issues seen in organizations with inadequate programs include:

- Selective destruction of documents—picking and choosing what should be kept or destroyed;
- Not having a written, rational plan for document retention and destruction;
- Keeping records too long;
- Failing to maintain records that are required by law or regulation;
- Destroying records (even using a written procedure) when government audits, regulatory action, or litigation is pending or imminent;
- Carrying out retention/destruction in nonroutine ways or that are not in the regular course of business.

Any of the above practices can cause significant regulatory and legal problems.

Knowing What Is Where: The Connection to Document Distribution

The better controlled your documentation distribution is, whether it be a push or a pull system, paper-based or electronic, the easier it will be to retain functional documents and properly destroy them when they are no longer needed. If the system is *not* controlled, not only will you have regulatory and performance problems at the point-of-use, but it will be very difficult getting rid of outdated copies.

It is important to specify the *copy of record* for each document type. For functional documents, this could be the "master" version that is produced by the documentation group or the department that is designated (by procedure) to prepare it. All other controlled copies of this document are simply copies that will be replaced as newer versions are released. Functional documents that are for reference only and go to one-time users ideally are destroyed after use, but more commonly stay in files until the files are purged.

The copy of record for records and reports that are required by the GMPs (e.g., cleaning logs and deviation investigation reports) also need to be identified. For most records, such as cleaning logs, this is simple: the original logbook or sheets are the copy of record. For laboratories, results that are placed in a lab notebook or on pre-numbered lab worksheets would be the copy of record.

The issue becomes more complex with reports such as deviation investigations. Different departments may create these depending on the firm's procedure. It is critical, then, that the copy of record is clearly known and designated. Different firms do this different ways. For example, some firms keep all the copies of record with the batch manufacturing documents. If you want to read the report, you need to go to the batch record. Other firms have deviation files in the quality control unit; various departments generate reports but the official report and related documents are kept by the quality unit. As with other GMP topics, create an SOP describing a process that works for *your* firm, and then follow it.

Documents that exist electronically also need to have a copy of record defined, particularly if periodic back-up tapes are made. This should be defined in your system's backup strategy.

Private or Personal Copies

A particularly difficult area to control is that of personal copies of documents, particularly of reports that someone has in his or her "private" files. Unless there is an active document management program in place, these records might be kept far beyond when they should have been properly purged and destroyed.

"Personal" files really are not private. Technically, if the documents and files were created by employees or agents working within the scope of their employment, the documents and files are considered documents of the organization. This extends to handwritten notes, "private" calendars, and diaries.

Knowing What Documents You Need and How Long They Are to Be Retained

The GMPs and other agency regulations are quite specific about documents that need to be produced and then retained; Chapter 3 discusses this in more detail. Regulatory agencies are also specific about retention times. Table 11.1 presents examples of these.

Table 11.1. Examples of Document Retention Times

Agency/Product Citation	Documents Covered	Retention Time
U.S. FDA—Drugs 21 CFR §211.180 (a), (b)	*Any production, control, or distribution document . . .*	Drug product *shall be retained at least 1 year after the expiration of the batch.* Certain OTC drug products lacking expiration dating . . . *for 3 years after distribution of the batch.*
	. . . all components, drug product containers, closures, and labeling . . .	Drug product . . . *at least 1 year after the expiration of the batch.* Certain OTC drug products lacking expiration dating . . . *3 years after distribution of the last lot of drug product incorporating the component or using the container, closure, or labeling.*
U.S. FDA—Medical Devices §21CFR 820.180 (Final Rule, Published Oct 7, 1996)	*All records required by this part . . .*	*. . . shall be retained for a period of time equivalent to the design and expected life of the device, but in no case less than 2 years from the date of release for commercial distribution by the manufacturer.*

(Continued on next page.)

(Continued from previous page.)

Canadian HPB - Drugs (4th Edition) C.02.021	*(1) . . . all records and evidence required to be kept or maintained . . .*	*. . . shall be retained for a period of at least one year after the expiration date on the label of the drug.*
	(2) All records and evidence required to be kept or maintained ...on the testing of raw materials and packaging materials . . .	*. . . shall be retained for a period of at least five years after the materials were last used in the production of a drug.*
European Community—Vol IV Good manufacturing practice for medicinal products (1992) 4.8	*The records . . .*	*. . . should be retained for at least one year after the expiry of the finished product.*
WHO—Drugs (1992) 14.8	*Records and assorted SOPs . . .*	*. . . should be retained for at least one year after the expiry of the finished product.*

Industry practice

While the regulations give the legal requirements, many firms have set their own internal requirements on the advice of their legal counsel. An informal survey of several dozen U.S. drug and device firms taken during workshops on the topic of documentation systems showed that most firms have GMP document retention times that go beyond the time required by the regulations. For example, many firms use a standard three, five, or seven years for batch records and seven years after the last use for functional documents like SOPs and specifications. A large number of firms have a simple practice: they keep master documents, records, and reports forever.

Determining your firm's policy for document retention should be done with a complete understanding of the legal implications (both benefits and risks). The policy should established in collaboration with legal counsel familiar with the potential issues.

Some of the points that should be considered in setting internal retention times are:

- The primary regulatory requirements for keeping the document (e.g., 21 CFR 211.180);
- Any other regulatory requirements that might apply (e.g., retaining contracts with governmental agencies);
- Any state or local requirements that might apply; and
- Any statute of limitations or limitations of actions (during which lawsuits or legal actions might be initiated).

Special documents

There are some GMP documents a firm produces that extend far beyond one lot or batch of product. For example, validation records covering the installation qualification (IQ), operational qualification (OQ), and performance qualification (PQ) of facilities and utilities. Since these systems impact each product made in that facility, they could be requested during an inspection at any time. Therefore, these documents, as well as the batch records of validation lots, need to be retained for at least one year beyond the expiry of the last lot of the product.

These documents should be specially marked and retained so they are easily and rapidly accessible to those needing them. These are typically accessed as reference documents much more frequently than typical batch records.

Retention Methods

With the knowledge of what documents need to be kept and for how long, you can consider the best, most cost-effective ways of retaining them. Retention methods need to address several areas, including what is actually stored (the original copy of record or a duplicate of the original), where and how it is stored, and the practices for accessing it.

Original copy of record

Many firms keep the copy of record in its original form: paper. For example, the batch record which would include the manufacturing instructions, release testing results, batch review checklists, etc. would be filed in a large folder. As long as all the information is readily available to a regulatory inspector, this method of retention is acceptable. Some firms make a photocopy back-up of all the documents and store them off-site in the event that the originals are destroyed.

In regard to e-mail information, there should be a standard way of collecting paper copies of important correspondence related to a lot, for example, messages concerning developments in a deviation investigation or testing failure.

The type of paper used is a concern with paper documents. For example, some lab instruments present their results on thermal paper, as do older types of fax machines. Heat, light, and time cause degradation of the images. Most firms photocopy these images onto more stable paper. If the paper images are to be retained for a long period, special, low-acid paper will help minimize deterioration.

Storage of paper documents should ideally be in a fire-resistant cabinet, rated for at least one hour. Alternatively, these documents can be housed in a special storage area that has a non-aqueous fire elimination system.

A difficulty of a paper system that has only one copy is the potential of loss of that paper due to misfiling, theft, or untimely destruction.

Duplicate Copy of the Original Copy of Record

Microform

The most widely used approach has been *microform*, where photographic images are taken of the documents and stored on rolls of *microfilm* or as small sheets of film known as *microfiche*. *Micrographics* refers to the entire process of recording images on film.

The process of microfilming images has been developed and evolved to a sophisticated state, with cameras and document feeders that can image both sides of a page simultaneously and at a very high resolution.

Source documents can be single sheets of paper or in bound notebooks. Also, microfilming of computer images or reports (computer output microfilm or COM) has been used for a number of years. (In this case, the microform version is in fact the original, since the image is created from the data and directly placed on the microfilm.) Polyester-based silver gelatin microfilm is estimated as being stable for more than 500 years. Other types of microfilm (e.g., diazo, vesicular) used for duplication of an original microfilm have a "shelf-life" of 100 years.

To access the image, microform readers are needed. These can be simple magnifying systems or more complex reader-printers. Since agency officials have the right to have copies of the documents, firms using microforms need to have the ability to make paper copies.

To make microform documents more accessible, computer-assisted retrieval (CAR) can be used. Here, an index of document names, keywords, dates, etc. and the corresponding image identifier is created using a database. This database can be searched, with the CAR system then accessing and displaying the particular image.

Digital Systems

Microform copies are photographic reproductions of the original; a microfilm reader is simply a magnifying device. On the other side, are the digital systems that store the information as a raster image (an electronic snapshot) or as a vector formula (as in a word-processing document). Each of these systems has advantages and disadvantages (Table 11.2).

One of the critical points in a digital system is where the document (or, more precisely, the data that represents the document) is stored. For long-term, archival purposes, the desired media is not magnetic (like diskettes or hard-drives) but an optical storage device such as a *Write-Once, Read Many* times (WORM) disk. Documents placed on a WORM disk cannot be removed or modified, something that helps ensure the trustworthiness of the document. WORM technology uses a laser to record (and later read) data onto the disk. CD-ROMs (Compact Disk-Read Only Memory) also use lasers to write onto the disk media. CD-ROM writing stations are becoming much less expensive, with desktop units less than US $ 1000.

Other storage media that have specialized uses include magneto-optical drives that are faster than CD-ROMs and WORMs, but are not methods that permanently write the information in a way that cannot be deleted or changed. This medium is useful for retaining large files of data such as digitized video.

Magnetic media, such as DAT (Digital Audio Tape) are still used frequently for routinely backing up a system. As with magneto-optical media, this is not currently an accepted way for long-term archiving of electronic documents.

There are some limitations in using electronic formats for archiving documents; there are several different issues. First, there can be compatibility problems—will Microsoft's Word 20.1 be able to "read" the SOP created in Word 7.0? (Based on experience to date, it is very doubtful.) There will need to be hardware and software that allow documents to be read.

The second issue is with the media itself. Research indicates that the anticipated life for optical disks and tapes is less than 30 years. While this may be acceptable for typical batch records, it may not be adequate for saving images of laboratory notebooks.

Table 11.2. Comparison of Raster and Vector Formats

Raster ("images")	Vector ("text")
• Image-based • Can be indexed with keywords • Can be overlaid and annotated • Cannot be easily searched (unless indexed with keywords) • Like electronic microfilm • Can be converted to text files using optical character recognition software	• Not simply an "image" • Can be indexed with keywords • Text, libraries can be searched using full-text search engines • Controlled change is easy • Can import documents from "native" programs such as MS Word, etc. OR the output of optical character recognition software

FDA's Electronic Records Regulation

One of the issues addressed in the FDA's Electronic Records; Electronic Signature Final Rule (FDA, 1997) is that of readability of electronic documents. In summary, if a firm uses electronic documents to meet any part of the FDA's regulations related to the Food, Drug, and Cosmetic Act or the Public Health Service Act, the system must be able to

> generate accurate and complete copies of records in both human readable and electronic form suitable for inspection, review, and copying by the agency.

In addition, the system must be able to protect records "to enable their accurate and ready retrieval throughout the records retention period." Chapter 2 discusses other aspects of this regulation.

Duplicates vs. Originals: What Must Be Kept?

When either microform or digital image reproductions are made from paper originals, what can be done with the paper originals? Can they be destroyed or must they be retained? Many firms are facing this question as they move into electronic documentation programs.

There are three issues involved that make answering this question difficult. First, consider why the FDA (and other entities) have historically placed more confidence in paper-based records:

> The traditional paper record in comparison, is generally a durable unitized representation [of an event, decision, etc.] that is fixed in time and space. Information that is recorded directly in a manner that does not require an intermediate means of interpretation . . . . Although paper records may be falsified, it is relatively difficult (in comparison to falsification of electronic records) to do so in an undetectable manner. In the case of paper records that have been falsified, a body of evidence exists that can help prove that the records had been changed; comparable methods to detect falsification of electronic records have yet to be fully developed (FDA 1997, p. 13433).

Second, regulatory agencies and the judicial system are slow to keep up with technological changes: industry met with the FDA in 1991 to discuss electronic records, more than six years before the Electronic Signature regulation was finalized (Motise 1992).

Third, there are two sections of the U. S. Code, "Uniform Photographic Copies of Business and Public Records as Evidence Act (UPA)" (28 USC 1732) and

"Uniform Rules of Evidence (URE)" (Rules §§ 1001-1003) that address copies in terms of evidence. Skrupsky (1991, pp. 152-163) discusses this much more fully, but in summary:

- The UPA applies only to microfilm or other duplicates prepared in the regular course of doing business. In other words, these duplicates need to be made as part of a routine program.
- Only one of two requirements must be met for microfilm or duplicate records: (1) the duplicates must accurately reproduce the original OR (2) the duplicates must form a durable medium for reproducing the original.
- The UPA permits the original record to be destroyed in the regular course of business ("unless its preservation is required by law").
- Under the UPA, duplicates can be used in place of the originals as evidence in administrative and judicial proceedings.
- The UPA also permits the use of duplicates of a duplicate (second generation microfilm) as evidence.
- The URE enables microfilm to be admitted as evidence unless (1) a question is raised as to the authenticity or the continuing effectiveness of the original or (2) it would be unfair to admit the duplicate in lieu of the original.

Ways of helping to assure the reliability and authenticity of copies is discussed below.

The FDA's view

According to the 21 CFR § 112(a), for records that are maintained but not submitted to the agency, persons may use electronic records in lieu of paper records as long as the requirements of 21 CFR 11 are met. In this case, paper records need not be retained.

Storage of the Retained Documents: Where and How

Physically storing documents is where direct and indirect costs can become significant. An 8 1/2 by 11 inch paper batch record for a simple tableted product can easily be four to six inches thick; batch records for parenteral products can be 12-18 inches thick when all the cleaning and sterilization records are included. If a manufacturing site is producing 25 batches of oral products per week, over a three-year period (assuming a two-year expiry) this adds up to nearly 2000 linear feet of paper in batch records alone.

Costs associated with storage include the space charges (i.e., rent), fire-resistant file cabinets or boxes, security, fire protection, etc. Packing, handling, organizing, retrieving, refiling, tracking, etc. add to the costs. Many firms who do not have the storage space or personnel will contract to outside organizations to handle document storage.

Reducing the amount of storage space required can reduce the costs, but the reduction process and media (microfilming or digital scanning) need to be considered in the equation. Another element is the potential loss of the documents and what that means to the organization. Duplicate photocopies are one way of backing up the information, but that is very costly (in material and storage space) compared to duplicating and storing a roll of microfilm or a CD-ROM.

The space for storing documents needs to be adequate and safe. While basement locations are good for some reasons (e.g., supporting all the weight of the paper and file cabinets), low-lying areas can be subject to flooding brought on by spring-time thaws or hurricanes. Potential spaces under high water-usage areas (e.g., kitchens, glassware washing rooms, autoclaves) should be avoided.

Storage cabinets and shelving should be raised at least six inches from the floor to prevent damage from small leaks, activation of the sprinkler heads, and floods. Drains, equipped with backflow prevention protection, should be capable of handling the output from the sprinkler heads.

The room should be tall enough so documents can be stored at least 18 inches from any piping, HVAC ducts, or electrical lines. Local fire codes and insurance requirements should be reviewed. More details for the design and construction of a documents storage area can be found in the National Fire Protection Association (NFPA) document entitled, "Protection of Records." (See the References and Readings section.)

Storage controls

In preparing to do in-house workshops on the topic of documentation systems, clients have given me tours of their document operations and archival areas. A surprising number of firms (in the U.S., Puerto Rico, and Canada) have extremely lax controls over their batch records. Even as a visitor, I have been able to pull batch records off the shelves and go through the material without any questions asked of me. The documentation clerk at one firm even suggested that I take a file back to my hotel and bring it back when finished.

What would happen if your firm lost a page from the batch records or all the documents contained in a batch record? How would you be able to reconstruct the manufacturing, control, and release process 14 months after it had been produced and released?

Wherever GMP and other important documents are retained, there *must* be stringent controls over who can access the materials and storage area, where the documents can be reviewed, and how they are refiled. This is important no matter what form the documents (or images) are in, but particularly when there is only one copy of record.

When documents are placed in storage, they need to be labeled so they can be easily retrieved when required. Typically all documents in a particular series (e.g., batch records, SOPs, specifications) are stored together for ease of accessibility and destruction.

Destruction of Documents

When documents, records, and reports have been used and their life comes to an end, they need to be destroyed as part of the document management program.

Periodically, one reads of firms (or politicians) getting in trouble for destroying documents, yet, that is what document management programs advocate. Confused? There are some important differences that distinguish a routine, acceptable document destruction program from one that gets people in trouble.

First, an acceptable document destruction program is defined by procedure. This procedure should successfully *exclude* from destruction records that have particular historical value or may be the subject of regulatory or legal action. Second, the program needs to be done as part of doing business. This should be as routine as closing the books at the end of a fiscal year. The more 'normal' and routine it looks, the fewer concerns it will raise. Third, the destruction program should generate records listing what documents were destroyed, when, and by whom. (Firms with well-established document management programs keep destruction records 7–10 years.)

In summary, you want to show that destruction of documents is a normal, well-controlled process and not one that is haphazard or could be thought of in any way as being done in bad faith.

Copies of documents that have been replaced because of revisions or that are no longer needed should be disposed of in secure "destruction only" containers or using a shredder. Most firms have containers or shredders strategically located throughout the facility. As long as these are not the copies of record, there is no need to record the destruction.

Documents that are being destroyed as part of the document management program can be destroyed by way of shredding or using a bonded document destruction service. The destruction of these documents should be recorded.

If the documents are on microform, the microfilm can be burned, shredded, or the emulsion can be chemically stripped. Local environmental laws should be considered if burning is used. A witnessed burning of film would be the most secure method of disposing of confidential information.

Movies and television programs are not exaggerating when they show people using tape and paste in reassembling shredded documents. All it takes is patience and determination. Biotech firms can be especially at risk of competitors going through their trash. The thinner the resulting strip from the shredder is, the more secure the destruction will be. Also, there are some shredders that cut the shreds into small pieces and some that destroy the originals by punching holes in the documents. Bonded commercial firms have large-scale units mounted on the back of trucks that grind documents into a pulp right at the client's site.

Destruction of Electronic Files

If you delete a document from a computer using the standard delete command or dragging the document to the Trash icon, the computer deletes that document from its directory, allowing that space on the hard drive to be overwritten sometime in the future. This means that the information can be recovered *unless* it has been completely overwritten. On the positive side, this allows a user to recover a file deleted in error using a recovery utility program. On the negative side, someone else, with malicious intentions, can do the same thing. There are special application programs that do "electronic shredding" of documents, assuring that all traces of the document are removed.

Your Document Management Procedure

The procedure you have covering document retention and destruction should include specific details related to the following points:

- Identifying the custodian who keeps the copy of record for GMP-related documents;
- When and how copies of documents (e.g., distribution copies, one-time use) are disposed of;

- When and how documents (copies of record) change from active to inactive status;
- A rationale for selecting and permanently maintaining certain documents that are historically important;
- How copies of record are to be maintained and where;
- Ways that copies will be authenticated during their preparation;
- What happens to the original documents if microform or digital image copies are made;
- How documents that are archived can be accessed for review;
- Ways to put document destruction on hold in case of litigation;
- The rationale for document destruction timeframes;
- The process for document destruction, including records that will be maintained;
- Training requirements for those involved in document retention/destruction; and
- Auditing retention/destruction activities.

Conclusion

If information is one of the most important resources a firm has, then thought and care must be given to how that resource is retained, protected, and used. Too little control and the information might never be seen again; too much control and potential users may not be able to access the information when it is needed. The costs involved for maintenance of the archived data, as well as the costs of not having the proper amount of information quickly available, need to be weighed very carefully so the proper balance is found for your organization.

Frequently Asked Questions

Q. What is the difference between copying a record and transcribing it from one place to another?

A. Copying is making a true, identical replica by way of a photographic technique (microfilming or photocopies) or by using a digital scanning tool. In this way, the original and the copy are considered identical. Transcription involves, generally, the handcopying or hand transferring of information from one document to another, such as when data from a certificate of analysis are transcribed from a vendor's form to an internal company document. It could also be information on a paper document entered into a computer's database. If a transcription is done, this must be noted. It should also be verified, whether the source document is retained or destroyed. Auditors and agency inspectors must have confidence that the resulting transcription contains *all* of the information, not a selection of it.

Q. *After reading about the legal implications of a document retention and destruction program, I am very nervous about setting one up. What if we do something wrong unintentionally?*

A. This book is meant only to discuss the issues and give general guidance. You are right . . . setting up a quality, defendable retention and destruction program is very important. It may be very useful to call in a professional (either an attorney with expertise in document retention systems or a person who is a Certified Records Manager) to review your system and planned procedure.

Q. *Our problem is that managers and technical personnel keep documents in their files forever . . . . I saw one memo on a stability failure from 25 years back. What can we do to get our system under control?*

A. All personnel need to understand the benefits and risks of documents and a systematic way of getting rid of them. Part of the solution is an education process. Another part is management's commitment in routinely using the process. The process must be a part of normal business operations if the retention/destruction program is to be credible to agency personnel and the legal system. For a program to be successful, *everyone* must be involved. Document audits should also examine the practice periodically.

Q. *A manager recently went ballistic when he found out that some documents had been destroyed that he needed. They were done according to the approved destruction schedule. Were we wrong in doing this?*

A. As mentioned above, education is needed for all people in the organization. If you were following an approved procedure and documented what you did, etc., technically you are on solid ground.

You may, however, want to examine your system to determine if problems like this might be avoided in the future. For example, are all appropriate people informed before destruction will take place, for example, in a memo that states what document series will be destroyed? This way, people have one final chance for exempting something for a particular reason. (This needs to be documented with a rationale.) Also, is the shelf-life for this document appropriate, i.e., does this series of documents need to be available longer than it currently is? One-off exceptions are dangerous . . . it can give people the idea that your firm is arbitrarily applying retention and destruction requirements.

Q. *Our training program uses sign-in sheets that we then manually enter into a database so we can keep track of things better. Do we need to save the original sign-in sheets?*

A. According to informal conversations with FDA experts, they really are not excited, in a positive way, when they see this type of situation. Their concern is that people can make mistakes. Also, if someone signs that they were present, that signature carries more legal (and regulatory) weight than someone's entry into a database.

One solution is to enter the data into your database that you can track and manipulate, but keep the paper document (or scan it as an image onto a WORM disk), which can be used to support any challenges that might be made.

An even better solution is be to collect electronic signatures of the attendees right on the spot using a PC or data logger.

Q. *How long should we retain our training records that show an employee was trained on the GMPs and the SOPs he or she uses?*

A. There are two typical situations when an auditor or agency inspector would want to see training records. In the first, he or she will determine if your firm has an adequate, ongoing training program. Having records to show that the employees are up-to-date in their training and that the training efforts have been in place for a number of years would be useful. You should have records to show that the employee has been trained on *all* of the SOPs he or she uses and is qualified to do the job. If you can do this, keeping records for 5-7 years may be adequate.

The second broad question that can bring up training records is for a lot that is "in trouble," i.e., there have been complaints or the lot is subject to a recall/withdrawal. Here, you would want to be able to identify people working on that particular batch and determine if they were qualified at the time they were making the product. In this case, you may need to review training records several years previous to when the batch was made, since not all training would have occurred in that year or the previous year. If you have products with a long expiry date (e.g., five years) you may need to keep your training records for 7-10 years.

Beyond the GMP issues, you should consider other regulations covering employment records.

It's spring-cleaning time, and the folks at Amgen, Inc., are throwing a Trash Bash, a mandatory cleaning of every file in the office, from manila folders to floppy disks. . . . "Most people are information packrats," says Sharon Hyder, a records specialist who helped Amgen organize the Trash Bash. And the predilection to hoard is exceeded only the ease with which computer users can copy and store data. "People are scared to death of losing anything, so they're constantly making copies and backups," says Aldis Porietis, a computer consultant.

Amgen's Mr. Hardy tells workers that as long as they follow the company's retention policies, management won't hold them accountable for discarded documents. . . .

Culling file cabinets proves easy enough. "It's very cathartic," says Mr. Hardy. By day's end, the department's 250 employees have filled 100 "Insta-shred" containers with 20,000 pounds of purged documents, or about 80 pounds per employee.

Wall St. Journal, May 22, 1997

Chapter 12

Preparing For and Recovering From a Disaster

Goals

- Define *disaster planning and recovery*.
- Discuss what to include in your disaster plan.
- Describe what should be included in a disaster recovery kit.
- Identify initial steps to take to secure documents after a disaster.
- Identify further steps to take in recovering from a disaster.

An airline pilot remarked that there are few real emergencies when flying a jet. If an engine catches fire, the crew follows a procedure that they have practiced in a simulator. If fuel unexpectedly runs low, there are plans in place for addressing the situation.

For documentation professionals there should also be few emergencies: there should be written plans that have been discussed and practiced in order to minimize the impact of a disaster, whether it be a fire, hurricane, flood, catastrophic systems failure, or some other act of God.

Having a Plan

You have heard the maxim that an ounce of prevention is worth a pound of cure. When it comes to disasters, a milligram of planning is worth a kilo of cure. If you have ever had a printing firm lose an electronic file or someone not keep their files up-to-date, you know the frustration, anger, time, and expense that go into recreating the information. What is even more astounding is that people who have been through this forget over time and get sloppy about backups and preventing disasters from recurring.

From a GMP point of view, your disaster plan should cover:

- Document-processing systems being used to create new documents;
- In-process manufacturing records;
- Electronic masters of all functional documents;
- Paper masters of all functional documents;
- Paper-based originals/copies of record;
- Electronic copies of record;
- Laboratory notebooks;
- Paper-based archive documents;
- Electronic archives;
- Microform archives; and
- Other important documents, records, reports.

(While you are preparing the disaster plan, you should also consider the retained ["house"] samples and stability samples.)

The Loss-Prevention Plan

The first plan your firm should have is for routine backups of existing information. This should be done in conjunction with the computer system experts in your organization or with a consultant. The process can be as simple as backing up files each night using diskettes or a tape drive (e.g., Iomega's Zip™ or Jazz™ drives) and, just as importantly, storing a recent copy of the back-up off-site. Having these backups will minimize the disruption resulting from a disaster and help you rapidly return to normal operations.

Is off-site storage of back-ups important? It can be. Fires, explosions, and other strange things unexpectedly happen. It is your decision on how prepared you want to be. *Not* taking any action is a decision as well.

Once you have the prevention plan in place, you can address how you can recover if a disaster does occur.

Disaster Recovery Plan

Disaster recovery plans consider the disasters most likely to occur at the particular location. For firms in Puerto Rico, this might be hurricanes and floods; for firms near or on fault lines, earthquakes are possible. (Several weeks after I taught a course on documentation systems, the Northridge, California, earthquake hit; the firm of one of the participants was severely damaged.) In the U.S. Midwest, floods and tornadoes are common. Fire and resulting smoke and water damage should be considered in all settings.

The plan should address two different sets of activity. First, *what can be done to get the organization back in business as quickly as possible*? Here, recovering the vital records (those documents essential to the survival of the firm) and getting them back into service are the primary issues. To do this, you will need to identify what your organization considers vital documents.

Second, the plan needs to address *recovery of other documents*, such as those that have been archived for regulatory or historical purposes. Your firm will probably have many more of these documents than those in the "vital documents" category.

Your disaster recovery plan should identify the roles and responsibilities of people who will be involved. What are the documents for which they are responsible? What are the people supposed to do? What do you want to be sure they do *not* do?

It is useful to break down the plan by types of documents and also by time. For instance, what can be done within an hour of finding wet documents is different from what you need to do if a document has been wet for 60 hours.

Making initial contacts

One of the elements in your plan is to know who to contact for specialized services. For example, you should have the names, addresses, and emergency phone numbers of services that can supply expertise and equipment. Your firm may want to have contracts in place with these vendors so that the response is not delayed.

Before a crisis hits, it is useful to contact firms that specialize in computer disaster recovery, particularly if there are systems that are not routinely backed up. Know their requirements for getting equipment sent to them. (Also, knowing the price of their services will be a stimulus to doing routine backups—a simple situation evaluation can cost $3–400 per machine; data recovery can cost up to $1500 for a standard PC.)

You may also want to identify mobile refrigeration services and industrial freezers and freeze-drying services. Be sure to record their emergency phone numbers.

Emergency Response and Salvage Wheel

A tool that will be extremely useful in creating your plan and then executing it is the *Emergency Response and Salvage Wheel* created by the U.S. Federal Emergency Management Agency, the National Endowment for the Humanities, the J. Paul Getty Conservation Institute, and others. This is a simple, unique job aide that gives specific instructions on how to stabilize a variety of documents that have been exposed to water. The Wheel is inexpensive (less than U.S. $10.00) and could be provided to all members of your firm's disaster team. (Details on how to order are provided in the Resources section.)

Practice using the plan

Once the plan has been written, it is useful to have a practice session. This could be a "talk-through" or a dry run (i.e., a simulation). A better way is to actually create a situation where people put the plan into action, perhaps by having them "save" old, outdated documents. This way, people can get a better feel of what they might encounter. Also, you may find parts of your plan that can be improved.

Emergency supplies

One other aspect of the plan is to have an emergency kit available with materials that would be used in the emergency response. Eulenberg (1986) recommends a list of supplies, including those shown in Table 12.1.

Table 12.1. Emergency Response Supplies (Eulenberg, 1986)

- Copy of recovery plan laminated in plastic sheets
- *The Emergency Response and Salvage Wheel*
- Interleaving supplies: at least 6 reams of 9x12 inch (new) newsprint or 10 rolls of paper towels, wrapped in plastic or florists (white) waxed paper
- 25–50 heavy-duty plastic bags, 30–40 gallon size, 2–4 mil thick
- 200 manila folders wrapped in plastic
- 5–10 black and red permanent marking pens, medium and wide tip
- 2 utility knives with retractable blades; extra blades
- 2 plastic letter openers
- 1 tablet of 9x12 mylar paper
- Crowbar, pliers, screwdrivers, Swiss Army knife
- 200 feet nylon cord
- Box of latex gloves*
- 6 pair of cotton (photographer's) gloves
- 10 dust/mist respirators (masks)
- 10 cotton rags/towels
- 2 high quality utility (kitchen) scissors
- 1 or more 30 gallon high-impact plastic garbage containers.
- Battery lantern, flashlights, batteries**
- AM/FM radio, batteries**

* replace at least once a year

** replace at least twice a year (place on preventative maintenance system)

The First Response

According to Eulenberg and other experts, the first step in dealing with water damaged documents, film, electronic media, etc. is to safely *get them out* from the disaster area. Know where the vital documents are and remove them first. Also, remove the wettest records, typically on the floor or under fire sprinkler heads. As the team works through the area, move on to damp records and then finally the dry ones, so as to prevent their getting damaged by humidity and mold.

For damaged computers, such as desktop PCs, the most important thing is to *not* try to restart them, as water from sprinkler systems and floods is filthy; the dirt can severely damage hard drives or other forms of magnetic media. Also, moisture levels can cause damage to the mylar matrix that forms the backbone of the magnetic media. Remove the equipment to a dry area and immediately contact a computer disaster recovery center.

An important point to remember is that "recovery" of documents does not mean that the recovered document will be restored back to the quality of its original state. The recovered documents will be harder to use and the image will be significantly poorer in quality than the original.

Working against time

There is a very real deadline your response team is working against. First, paper records tend to disintegrate after 48 hours of getting wet. Second, bacteria and molds will grow to high levels after this time point (especially in tropical and subtropical locations). For film (e.g, microfilm) or photographs, there is also a time pressure, both because of the damage water can do as well as the growth of molds and bacteria. Color photographs are particularly at risk.

If paper is only damp, the damage is very recent, or only a few pages are involved, natural airdrying or using blotting pages is an option. This can be started right away.

A tactic that will give you a little time (to a total of 48 hours or so) to consider various options is to keep wet things wet. For instance paper, books, microfilm, and black and white photos should not be simply dried, as they can be damaged beyond repair. In the case of microfilm, allowing the film to dry, even partially, can cause the layers to stick together and potentially destroy the images.

To recover or not to recover . . .

That is the question you will need to ask. If there are other backups available (e.g., kept off-site or in another medium), the decision may be made to not recover the documents. In some situations, there may be other documents available that could be used to recreate the documents. This can be a legal risk that should be discussed with management, legal, and, possibly, regulatory officials.

Paper on ice: freezing

One of the best ways of minimizing further damage to paper documents is to freeze them. This does several things. First, it prevents additional decomposition of the paper fibers and stops the growth of microorganisms. Second, if the documents are to be freeze-dried, freezing is a prerequisite step. Third, freezing gives you time to think about what should be done next.

If the documents are archived for regulatory reasons, e.g., batch records of a product made two years ago, and you need to simply keep the documents another two years to comply with the regulations, you may choose to keep them frozen. If, for some reason, you need to access them, you can freeze dry them at that time. Yes, this may delay the response to a regulatory question, but it would be considered reasonable if the disaster had been properly documented. Also, recovering documents on an as-needed basis would significantly reduce the costs of recovery.

Recovery of microfilm

Eastman Kodak, one of the world's largest manufacturers and processors of microfilm, recommends that "if there are no local facilities for rewashing and drying the films immediately, promptly place the films in a water-filled container and send them to a laboratory where they can be washed and dried properly (Kodak, 1996).

Firms that purchase Kodak microfilm and have it processed by a Kodak-authorized processor that subscribes to the Kodak Image Guard Program can utilize the services of Kodak's disaster recovery unit. Kodak will rewash, dry, and return rolls of film to the client without charge. General information can be obtained from the local Kodak Business Imaging Systems office. Specific information can be obtained from the Kodak Quality Control Laboratory in the United States at 1-800-352-8378.

Recovery of damaged photographs

Eastman Kodak recommends the following for photographic materials damaged by water:

> First and foremost, it is important that these materials (negatives, slides and prints) NOT dry out prior to separation

and cleaning. If they do, they may "block" (stick together), making subsequent separation more difficult or impossible. On the other hand, materials allowed to stay wet for extended periods of time become very fragile. Accordingly, separate and clean your materials as soon as possible. As a precautionary measure, wear rubber gloves while working with them—contaminants or biological growths may be present.

If thorough cleaning is not immediately possible, rinse your materials in a tank of cold clear running water until the water overflow runs clean. This is to remove any loose dirt and debris. Do not run water directly on them, as this may cause physical damage to the already-softened photographic emulsions. Keep them in a container of cold, clean tap water. The cold temperature minimizes further softening, so place the container in your refrigerator or use ice to keep the water cold —the colder the better. The chlorine in your tap water will prevent biological growth. However, keep in mind that chlorine dissipates within 24 to 48 hours, so change the water every day. Do NOT add chlorine to the water—higher concentrations can dissolve photographic emulsions.

As soon as possible, work on the photographic materials. Start by removing the smallest quantity possible from the storage container, and work with them submerged in a sink of clean water. Gently separate the films or prints from each other or their storage material. Do not force the separation—you may cause further damage. Separate them as much as possible, then return them to the water bath while you start another batch. Periodically return to previously worked batches to do additional separation. Eventually, you may have no choice but to force the separation and accept the resulting damage.

Once separated, further cleaning may be needed. First, wash your materials in cold running water to remove any loose substances. Remember to keep them submerged, since dried-on debris is difficult or impossible to remove. If necessary, use a tuft of cotton or a soft foam rubber brush to dislodge stubborn debris. Finally, rinse the materials one last time and hang them (unrolled, where applicable) on a line to dry. Prints should be simply hung to dry. The use of KODAK

PROFESSIONAL PHOTO-FLO Solution on black-and-white negatives and slides made on KODACHROME Film, or a stabilizer on color negatives and slides made on KODAK EKTACHROME Film, will facilitate cleaner and more uniform drying. PHOTO-FLO and stabilizers are available through photographic dealers and/or photofinishing labs.

After all efforts to salvage the materials, you can consider additional restoration. Reprinting negatives or making copies of prints would be the first step. Further restoration may be possible through retouching and then recopying. Today, digital imaging can offer significant benefits in restoration. While the costs may still be relatively high, these services are becoming more readily available and the costs are diminishing (Kodak, 1997, reprinted with permission).

Documenting the Disaster

The disaster plan should include provisions for documenting what occurred, the response, and the location and physical state/location of the damaged records. All documents being restored or that are frozen should be known; if one of these is requested, you should explain to the inspector that it will take additional time to access it. A well-organized file with photos and memos that can be presented to the auditor or inspector should be part of the explanation.

Conclusion

Preparing for a disaster is essential for a speedy recovery. People at all levels should know what to do, what documents are vital, and the what steps should be taken to minimize the damage and begin the recovery process. Backups of information, stored off-site are one of the best ways of quickly getting back into business.

Filed and Forgotten

During World War II, with their supplies of oil cut off, the Germans, after years of effort, developed a way to derive gasoline from coal. In fact, from 1944 until the end of the war Nazi tanks ran almost exclusively on coal-derived gasoline.

In 1975, at the height of the oil crisis, a chemical engineer at Texas A & M University, along with a colleague, decided to look at captured German documents to see what they could learn about the Germans' synthetic oil program. The two found that voluminous records had been filed and forgotten. Some were at the National Archives in Washington. Others were stuffed into crates in government buildings around the country. Even though the war and subsequent events in the Mideast showed the importance of synthetic fuel technology, until the two professors came along, the documents had lain untouched in government files for thirty years (Goldberg, 1984).

CHAPTER 13

Functional Documents

Goals

- Define *functional documents*.
- Provide examples of what functional documents are.
- Describe what functional documents need to include.
- Discuss ways of making functional documents more useful.
- Provide examples of some typical functional documents.

Functional documents are documents that prescribe, proscribe, or specify. In other words, these documents tell someone what to do, what not to do, or what is expected or required. People use them as they perform an activity, make a decision, or as a reference or training tool. Functional documents are usually used repeatedly, for example, lab methods that are used over and over in performing an assay.

Functional documents are produced as defined by procedure, encompassing the generation, review, approval, etc. steps described in earlier chapters. They include documents such as the following:

- Policies,
- Standard operating procedures (SOPs),
- Laboratory methods,
- Guidelines,
- Protocols,
- Master formulae,
- Work instructions,
- Specifications, and
- Drawings.

Let's examine some specific functional documents to see when they are used and details about what they would include. An emphasis will be placed on procedures as they are particularly and universally important.

Policies

Policies are documents used to influence and dictate actions and decisions. They tend to be broad in scope, affecting many different organizational, business, or geographical units. For example, an organization may have a corporate quality policy manual that applies to all of its sites. All sites and organizational units take direction from the quality policy manual and create their own local policies or procedures that fall within the broader corporate policy.

Policies describe primarily the "what" and, to a lesser extent, the "who" and the "whys." They *do not* provide the specifics of how to accomplish a task; procedures, methods, and work instructions do that.

Organizations wanting to be certified under ISO 9000 are required to have a quality manual "to define an outline structure of the quality system while serving as a permanent reference in the implementation and maintenance of that system" (ISO-9004, 5.3.2.2, 1994). Implementation of the policies is defined by procedures. Figure 13.1 shows the typical relationship between policies, procedures, work instructions, and records. Figure 13.2 provides an example of a policy on auditing.

What should be included in a policy

Policies should include the following elements, some of which can be built into the standard format of the document. Those items marked with an asterisk (*) are most important.

- Policy title*
- Control number*
- Effective date*
- Approval signature, stamp, or statement*
- Scope (to whom, when the policy applies)*
- Policy content*
- Organization name
- Summary

Figure 13.1. The ISO documentation system.

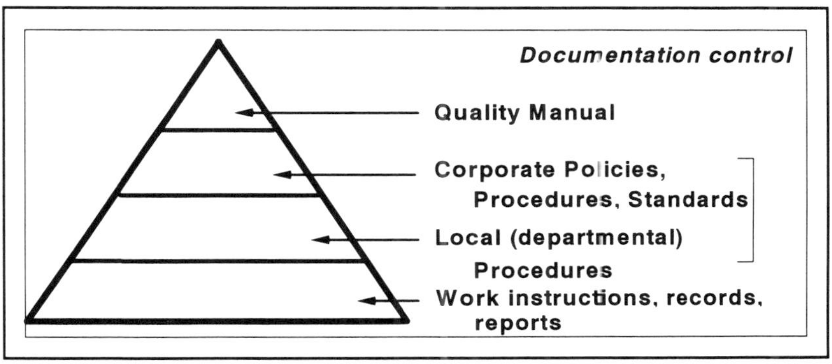

Figure 13.2. Example of a policy (Vesper 1997).

OMEGA HEALTH CORP.	Title:	**GMP Auditing**
	ID No.	QP101.2
Quality Policy	Effective:	Oct 1/96
	Approval:	Qcom [Approval Code] Aug 96

Goal	Periodically evaluate facilities to determine their success at meeting local regulatory standards (GMPs) and corporate quality requirements.
Scope	This Quality Policy applies to all OHP companies and subsidiaries worldwide.
Policy	1. OHP Corporate Quality Assurance (CQA) is responsible for establishing a program to systematically evaluate all GMP facilities on a regular basis. 2. Expectations, standards, and references will be developed, published, and distributed by CQA to sites at least six months prior to the scheduled audit date. 3. Each year, in the fourth quarter, CQA will develop a Proposed Audit Schedule for the upcoming year. The corporate Quality Committee will discuss and approve this at their Nov. or Dec. meeting. 4. Audit dates must be agreed to by site management and CQA.

(Continued on next page.)

(Continued from previous page.)

5. CQA will lead all audits and include at least one auditor from Technical Operations.

6. At the end of each GMP Audit, the audit team will provide site management with a preliminary oral report of their findings and its significance.

7. Draft reports will be reviewed and commented upon by CQA and the audited site's director.

8. Completed audit reports will be provided to the following within 20 business days following the audit.
   - Site Tech Ops director
   - Site Quality director
   - Director of CQA
   - VP of Quality
   - Executive VP of Operations

9. When there are audit findings requiring corrective actions, these actions shall be documented and approved by the site directors within 20 business days. Time frames for corrective actions are permitted. Action plans are to be addressed to the Director of CQA.

10. CQA will follow-up on the corrective actions. When completed, CQA will publish a Audit Closure Memo.

11. The Quality Committee will monitor CQA auditing activities and site corrective action plans.

Standard Operating Procedures (SOPs)

SOPs or simply "procedures" are functional documents that are instructions on how to perform a task (e.g., disassembly and cleaning of a mixer) or the process for making a decision (e.g., batch record review). The procedure breaks the task down into steps and substeps; SOPs may or may not contain the level of detail found in job instructions (see below).

What should be included in an SOP

The elements found in an SOP vary from organization to organization, as do the sequence of the elements within the SOP. Typical elements are listed below; those that are essential are marked with an asterisk (*).

- Procedure title*
- Control number*
- Effective date*
- Approval signature, stamp, or statement*
- Scope (to whom, when the procedure applies)*
- Procedural content*
- Organization, department name
- Summary
- Previous version
- Reasons for revision
- Approval date
- Next review date/expiry date
- Definitions, acronyms used in the SOP
- Key words
- Example of forms required by SOP
- Diagrams
- Flowchart of process
- Writer name, department, title

Level of detail

Procedures can be written to various levels of detail, as shown in Figure 13.3. For example, the word *wash* is a '"macro" word—a verb that encompasses a number of substeps that are more fully listed and described as the level of detail increases.

Choosing the proper level depends on a variety of factors, including how frequently the task is performed, its criticality, the amount of training provided, etc. Table 13.1 provides guidance on when lesser and higher amounts of detail are appropriate. For procedures used in drug and device manufacturing, high and fine levels of detail would be expected for most tasks.

Figure 13.3. Levels of detail found in procedures.

Low	Wash all stopper containers
Medium	Wash all stopper containers in the MDT cabinet washer using washer setting #3.
High	1) Load all stopper containers into washer trolleys with the open-side facing down. 2) Turn cycle dial to washer setting #3.
Fine	1) Load all stopper containers into washer trolleys with the open-side facing down according to diagram. 2) Turn cycle dial to washer setting #3 on control panel. 3) Wait for green "Wash Ready" light. 4) Press START button to begin the washing cycle.

Table 13.1. Relationships Regarding Level of Detail

Less detail *is appropriate when:*	***More detail*** *is appropriate when:*
• Task is frequently performed.	• Task is performed only occasionally or in an emergency.
• The same people perform the task.	• Different people are involved (e.g., rapid turnover of workforce).
• Variation of performance is allowed.	• Sequence of steps, degree of accurate performance is important.
• There is ample time to practice.	• Little time is available to practice.
• Outcome of task is not critical (e.g., trash removal in office setting).	• Outcome of task is extremely critical (e.g., emergency response to a leaky solvent tank).

The W-H questions

Procedures should answer the W-H questions: who, what, when, how, etc. Specifically, the SOP should include:

- **What**–the task, steps, and substeps that need to be accomplished;
- **How**–the details of performing the actions;
- **When**–the sequence used or the cues given to prompt the actions;
- **Who**–this may be assumed to be one person, but specify others in more complex procedures; and
- **Where**–the specific location of where the action(s) are to be performed.

Determining what procedures are needed

A careful reading of the GMPs provides a *minimum* list—a starting point—for creating procedures that are necessary. Certain auditing programs, for instance, the DRUMBEAT approach (Fields, 1995), give guidance on what SOPs a firm should have.

Another method, which is more a rule-of-thumb for deciding when SOPs are needed, uses several simple questions. If the answer to any of them is yes or maybe, a procedure is called for. The questions that are asked are:

- Is the process/task important?
- Could the task, activity, decision, etc. potentially affect the safety, identity, strength, purity, quality, or fitness-for-use of the final product?
- Are various people involved in doing the activity?
- Is the activity done more than once?

Another concept to be aware of when determining what SOPs are needed is the *cascade effect*, an example of which is shown in Figure 13.4. This means that there are a number of functional documents needed to support one relatively "simple" SOP required by the GMPs.

Figure 13.4. The cascade effect of procedures.

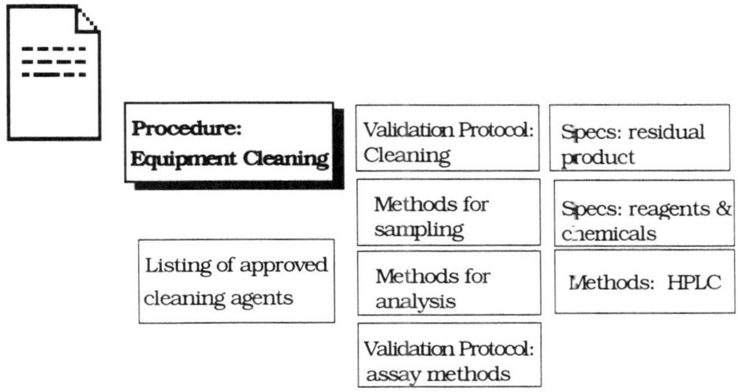

Defining what procedures are needed can be done according to the following steps:

1. Look at the particular activity or decision process.
2. Identify the task(s) that are needed to accomplish the activity or make the decision. For complex activities, there will be one procedure for each task. For simpler ones, or those that can be completed by one person in a unit of time, several tasks can be combined into one SOP.
3. For each task, create a unique, descriptive title.
4. For each task, write the purpose of the SOP, emphasizing the outcome or output of that task.

How "big" should an SOP be? How much should it cover? Should an activity be described in several, shorter SOPs or in one long document? Procedures should be created in "chunks" or informational units that directly relate to the task being performed. If a procedure is very large and covers everything related to the task, it will be difficult to use. People will have a hard time finding the information they need. For example, one drug company had a procedure describing their stability program that was 68 pages; no one, except the manager of the program, had read the entire document. Dividing the overall program into smaller, more focused SOPs makes the procedures easier to use. Also, shorter documents are easier to create, maintain, and revise.

Collecting the content for a procedure

One of the simplest ways of developing the content for a procedure is to "flowchart" the process or rationale used in decision making. The flowcharts do not need to be complex or sophisticated; they need to clearly show the steps, the sequence (if important), decisions made, and who is involved. A worksheet, such as that shown in Figure 13.5a-b, is useful in collecting other relevant information.

Some of the particular questions to ask as you prepare a procedure are:

- What is done, when, and by whom (refer to the "W-H" Questions discussed above)?
- What are the starting conditions and inputs (e.g., materials, information, documents)?
- What are the ending conditions and outputs (e.g., products, information, records)?
- What are the tools (physical and informational) used?
- How will the person know when they have succeeded in performing the task (e.g., what is the specification for the task)?

The Worst Printed Book

Nomination for the book with the most printing errors goes to a little book issued in 1561. Entitled *Missae ac Missalis Anatomia*, it runs 172 pages but contains 15 pages of errata. The author, trying to explain the march of misprints, declared that the Devil had made the printer do it: Satan, said the author, had soaked the manuscript to make it nearly illegible, thereby causing the printer to misread it and commit so many mistakes.

What is noteworthy about this fiasco is that the author did not blame the publisher, but blaming Satan was part of the erroneous views of the day about printing. It was felt printing was a "black art" and the work of the Devil since church and secular authorities believed that exposing the common populace to books would lead them astray (Goldberg, 1984).

Figure 13.5a. Example of a procedure development worksheet.

Procedure Development Worksheet

Process: ____________ Assigned #: ______

Task: ____________ Due date: ______

Contact: ____________

Flow diagram	Who	Cues: when/if	What	Spec	Com'nts, cautions

Figure 13.5b. Example of a completed worksheet (first page).

Procedure Development Worksheet

Process:	Failure Investigations	Assigned #:	QA732.2
Task:		Due date:	July 22/98
Contact:	Tara Contin		P 1 of 4

Flow diagram	Who	Cues: when/if	What	Spec	Com'nts, cautions
Prob. Observed	• Anybody	• Lab failure • Complaint • Process/non-proc. failure	• Problem is observed		
Failure Invest. process initiated log team hold notice	• QA Rep	•	• Assigns case# • Logs case#, description in failure log • ID's person or team to work on invest. • Lots involved put on hold	• Must be started ~~w/in~~ T.E. within 24 hrs of problem observation	• If serious/significant, QA mgr. is informed
Problem is defined	Investigator or team	1st step in investigation process	• Problem is defined • Problem statement created	• As defined in Inv. Handbook	

QA56.1

Providing guidance

Guidance is something that is extremely simple, but rarely provided in a procedure. Suppose you are in a new city, driving from the airport to someone's office for the first time. They have given you only a few instructions; as you are driving, you are wondering if you are getting close to the next turn you must make or if you have somehow missed the street and are now going too far. Had they provided you guidance, such as, "when you see the Amoco Gas Station, get in the right-hand lane, as you'll make a right turn at the next light," or, "if you see the train tracks, you've gone too far," you would be much more confident that you would arrive.

In SOPs, you can provide guidance as well. Some particular forms that are useful include:

- **When you are almost there:** "Before reaching the endpoint of the reaction, there will be a steep temperature rise, followed by a period with a rise of only 2–3 degrees C.
- **How long it takes someone to perform the task:** "Setting up the Bronski filler will take an experienced operator 3–4 hours."
- **When you have gone too far:** "If the bubbles in the liquid are no longer separate, but become a stream of air, you need to begin again."
- **When to call for help:** "Call Tech Support immediately if the monitor displays *System Error -190*."

What the SOP can look like

Procedures can take a variety of forms, but they tend to appear similar, primarily because most people write procedures to look like the ones they have used. In this way, the traditional format is perpetuated.

There is no one best way for presenting information in a procedure. The important point is how to present information in such a way that will *help* the user(s) do their job safely, effectively, and consistently. Some formats for the procedural description (the instruction section of a procedure), when they are best used, advantages, and disadvantages are discussed below.

Narrative

A narrative format, as shown in Figure 13.6a, has steps and substeps combined into paragraphs. This approach is (fortunately) not frequently seen in healthcare manufacturing, as it presents the instructions in a hard-to-follow manner. There are no advantages to this format for the user.

Outline

The outline style for a procedure is the one most commonly used. As shown in Figure 13.6b, the information is broken into steps and substeps. One difficulty with this

Figure 13.6a. Example of a narrative procedure.

5.0 Pressure to Input Calibration

After removing the transmitter from the loop by disconnecting the input and outputs, connect the pressure source to the transmitter input fitting. When a multimeter is connected to the output terminals, adjust the transmitter for the appropriate pressure to read 0% on the output and then record input pressure, current, equivalent, initial output current, and the tolerance code.

Figure 13.6b. Example of an outline style.

5. Pressure to Input Calibration
    5.1. Connect pressure source to the transmitter input fitting.
        5.1.1. Remove transmitter from the loop by disconnecting the input and outputs.
    5.2. Connect a multimeter to the output terminals.
    5.3. Adjust the transmitter for the appropriate pressure to read 0% on the output.
    5.4. Record input pressure, current equivalent, initial output current, tolerance code.

style is that some writers break down the actions into sub-sub-sub-substeps (e.g., 5.2.1.1.1), making the document difficult to follow. This is an adequate method of presenting information if the steps are properly written and there is an action verb (e.g., "connect," "remove," "adjust") at the start of each step/substep and the levels do not go beyond three or four deep.

Multi-column

The multi-column approach to procedures is extremely useful and easy to read. This format can help identify multiple performers or, as in Figure 13.6c, have the primary information on the left and supplemental information on the right. This way, experienced performers can just follow the steps on the left and novice performers can make use of the added detail on the right side. This format is very helpful when a wide variety of people are using a procedure and also to simplify training. A variation of this format (Figure 13.6d) includes drawings, especially useful if reading skills are low or if users do not have a common language.

Figure 13.6c. Example of a multi-column format.

5.1	Remove transmitter	• Disconnect input and output from the loop. • Input leads are white; output leads are black. • Do not cover lead ends.
5.2	Connect pressure source	• Use pressure source AGR-100 to connect pressure source to the transmitter input fitting.

Figure 13.6d. Example of a multi-column format with drawing.

5.0 Pressure to input calibration

5.1 Remove transmitter.	Disconnect input and output from the loop. Input leads are white; output leads are black. Do not cover lead ends.	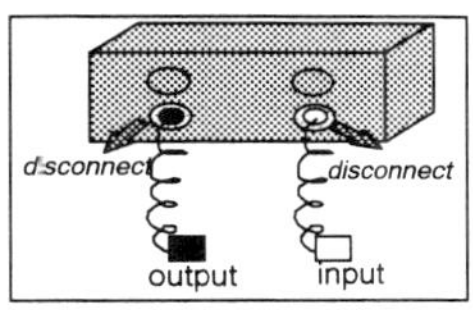
5.2 Connect pressure source.	Use pressure source AGR-100 to connect pressure source to the transmitter input fitting.	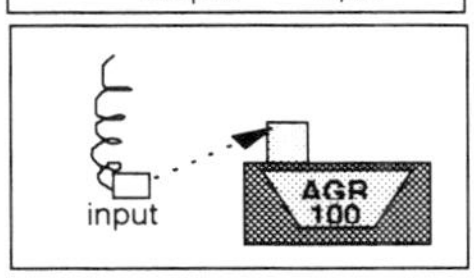

Decision tree

The procedure in the examples above are concerned with physical tasks—sometimes called motor skills. Another type of tasks involves decisions, such as the thought process in releasing materials or components for use in manufacturing. An example of a decision tree is shown in Figure 13.7. Some firms use decision trees as a way of documenting the decisions used for a particular process: users have a copy of the tree on which they trace the path used. Flow charts with decision points (typically diamond shapes) are a variation on this format.

Figure 13.7. Example of a procedure given as a decision tree.

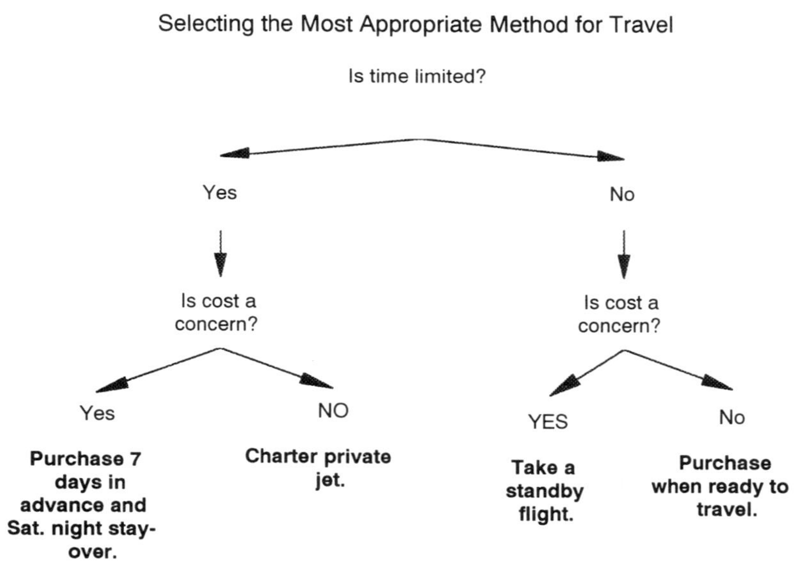

Writing conventions within SOPs

There are some standard ways of presenting information within an SOP. For example, numbering steps (1., 2., 3 . . .) means that the sequence presented is important and must be followed. If the sequence is not important, the actions can be presented as bullet points or with the statement, "steps 3 to 8 can be performed in any sequence."

If there are cautions or warnings to be included in the procedure, such as when adding potentially hazardous materials, these words should be used *before* the action is called for. These key words can be set in all caps for emphasis or surrounded with a box as shown in Figure 13.8. The term, *warning* implies that the outcome has a serious potential risk; *caution* is used when the outcome is not as serious.

Figure 13.8. Use of caution and warning statements.

CAUTION

Hydrochloric acid can cause damage to the skin, eyes, and lungs. Wear a full-face air-supplied respirator, apron, and rubber gloves when using this material.

WARNING

HIGH VOLTAGE !

USE LOCKOUT PROCEDURE BEFORE BEGINING DISASSEMBLY OF BLENDER

Methods

Methods are documents typically used in laboratories or when conducting in-process testing. They differ very little from general procedures, except that the content and its arrangement is optimized for laboratory use. Methods need to be prepared according to a procedure, reviewed, approved, and properly controlled.

Guidelines

Guidelines or guides are functional documents that provide additional information to the user. Unlike other functional documents, the user has some discretion in whether or not to follow the guideline. For example, guidelines might be useful in helping professionals determine how to select outside conferences that best meet their needs. A guideline should not be used if variation cannot be tolerated in the performance of the task.

"Master" Documents

Master documents include protocols, master batch records (or formulae), and device master records. These functional documents are used repeatedly as different batches or lots of items are produced or if an operational qualification is conducted on similar pieces of equipment.

A particular concern with master documents is that any mistake they contain is transmitted into the specific documents that are produced from them. For example, if a calculation is used to convert gallons into liters and the calculation is wrong, that error will be perpetuated.

For this reason, GMP regulations require that master documents, specifically master batch records and device records, be carefully checked and approved. Additionally, each time a batch-specific document is created from the master, the "daughter" document must be formally checked to be sure all pages are present, readable, and correctly identified.

Master documents must be prepared according to a procedure that includes all relevant GMP requirements. Also, they must be reviewed, approved, and controlled.

Protocols and master documents generally include references to other documents such as SOPs, methods, and specifications. Unless there is a specific reason to identify a particular version of these documents, it should be assumed that the most current one available at the point-of-use will be used.

Verification of critical steps

Master production records include spaces where those doing operations initial that they have done the step. For critical steps, there are places where a verifier or witness can initial, adding weight to the fact that the activity was properly performed. Thought must be given to how much added value each verification adds. One firm, in an effort to ensure things were done properly, added a verification line to each step. Soon, deviations started occurring because there were not enough people present to signoff as verifiers

What is critical? The U.S. drug cGMPs do not give a definition for this; however, guidance can be taken from other documents, such as the Canadian GMPs for drugs (HPB 1996), which define a critical process as *a process that may cause significant variation in the quality of the finished product*. Versions of the U.S. medical device

GMPs prior to the 1996 Quality System Regulation defined critical operation as *any operation in the manufacturing of a critical [medical] device which, if improperly performed, can be reasonably expected to . . . affect [the device's] safety and effectiveness* (§ 820.3(g)).

There are terms that are sometimes confused with *verify*. Firms or departments should define the terms *check*, *verify*, and *witness*. *Check* could mean that another person asked the operator if he or she did a certain task. *Verify* might mean that someone has some corroborating evidence to indicate that something was done, for instance, that properly completed raw material coupons and empty containers indicate that the raw materials were added to a tank. *Witness*, on the other hand, tends to mean that a person actually saw the event take place—the witness has personally experienced the action. Two other, similar words used in the GMPs that firms may want to define are *examine* and *inspect*.

Work Instructions

Work instructions are functional documents that are more specific and detailed than procedures or methods. When procedures are written at a low-level of detail, i.e., not very specific, work instructions may be more valuable. Batch manufacturing records that include detailed instructions are equivalent to work instructions.

If work instructions are used, they should be generated, reviewed, approved, controlled, and distributed as are other functional documents.

Drawings

Drawings can be used to specify and prescribe the arrangement or assembly of parts going into a medical device or to document the "as-built" situation—a record of what was done. In this chapter, we are considering documents that specify or prescribe.

Production or assembly drawings are more frequently used in medical device manufacturing than in drug production. They also tend to be one of the most difficult types of documents to keep up to date, in part because of the special skills and/or equipment required.

Specifications

Specifications are functional documents that define what is acceptable—they identify what is important about a particular object or material and also give the standards that an acceptable object or material must meet. The purpose of a specification is to communicate to the supplier, purchaser, and all potential users of the material what is "good." Specifications are also used in the decision making process. Here, people who must evaluate the product or material have a defined list of what is important to evaluate and what the acceptable values are. If the material meets these criteria, they can make the decision that it is acceptable; if it does not meet the criteria, it is not acceptable.

The more clearly and precisely a specification presents the information, the easier and more consistently decisions can be made. Considerations in preparing specifications are included below.

Types of specifications

People with an engineering background generally look at specifications as a much broader category of documents than do documentation specialists working in the drug industry. Table 13.2 identifies a number of specifications that a firm might have; several of these have different names that are more commonly used in the pharmaceutical industry.

Table 13.2. Examples of Specifications

Type of Specification	Description
Target ("design brief")	• Created at the beginning of the design process to define the design considerations of the object.
Functional	• Created during the design process, defining what the object should be able to do.
Product	• Parameters and criteria the item must conform to when it is completed.
Material	• Requirements of the components used in producing the object.
Process	• Details of the actions used to manufacture the object.
Inspection	• Parameters and criteria to be examined and determined to be adequate before release of the object.
Test	• Parameters and criteria to guide testing to determine adequacy at various stages of manufacture.

(Continued on next page.)

(Continued from previous page.)

Acceptance specs	• Criteria for accepting other types of specifications as adequate before approval or release of the object to a customer.
Packaging and labeling	• Parameters and criteria for all packaging components and labels, and how the object is to be labeled and packaged.
Installation	• Parameters and criteria recommended for the successful installation of the object.
Use	• Parameters and criteria recommended for the successful use of the object.
Maintenance and calibration	• Parameters and criteria recommended for the successful continued use of the object in accordance to its functional specifications.
Disposal	• Recommended methods for disposing of the object or waste streams produced while using the object.

What's included in a specification

Specifications need to be tailored not only to the organization, but to the item being defined in a particular document. In general, all specifications should have at least the following:

- Company name/division;
- Item name/title (i.e., what the spec is defining);
- Other item identifiers;
- Specification number;
- Specification revision number;
- Effective date of the specification;
- Approval signature(s), dates; and
- Details of the specification (parameters and criteria).

Frequently seen problems of specifications

According to McRobb (1989), there are three typical problems related to specifications. These problems involve the following:

- **Words**—using nonstandardized terms that are not clearly defined or using words inconsistently.
- **Concepts**—excessive and/or nonvalue adding requirements and excessive (i.e., poorly controlled) linking to other documents or information sources.
- **Presentation**—lack of understanding, poor layout of the information, and inadequate proofreading.

If careful thought is given to these in advance, not only will the specifications be easier to use, but decisions will be easier to make.

Creating Better Functional Documents

If you have considered the general points discussed earlier (Chapters 4, 6, and 7), you are well on your way to producing functional documents that will be useful tools. Some other, more specific guidelines include the following:

- **Use short sentences.** As people read, they tend to concentrate on the first few words and then may skip the remaining words in the sentence. Keeping sentences short and loading the important information at the beginning can help.
- **Use words appropriate to the audience.** Most operational documents should be written at the 6–8th grade levels; technical documents should be written at the 9–12th grade level. A non-U.S. firm did a reading analysis of its workforce and found its operational people to be reading English at a 4th grade level.
- **Keep the writing simple.** Don't overload a paragraph with multiple concepts. Use conjunctions that are simple to understand (e.g., *but*, *so*, *for*; not *however*, *therefore*).
- **Be terse.** Get rid of unneeded words. Instead of *write down the identification number of the tank* ___; say *Tank ID:* ___.
- **Use the active voice, not the passive**. For example, do not say *all raw material coupons should be attached*; say *attach all raw material coupons.*
- **Use positive statements** whenever possible. These both are easier to understand and create a better response in the user. For example, *use positive statements* instead of *do not use negative statements*.
- **Use specific ranges** whenever possible. If the process (and validation data) supports it, use a range, such as *mix for 9 to 11 minutes* instead of *mix for 10 minutes*. In the latter example, anything above or below the 10 minute point would be considered a deviation and must be docu-

mented. The more clearly the range is noted the better. For example:

9–11 minutes

is better (i.e., more clearly understood) than

10±1 minutes

and much better than

approximately 10 minutes.

- **Consider writing functional documents and training people to use as two connected parts with the same goal.** The goal is to get people to perform specific tasks or make decisions using the SOPs, specifications, etc. This unified concept will make for more useful documents with less training required. The results are more fully implemented documents and fewer deviations in less overall time.

Conclusion

Functional documents in an organization represent a huge investment in how the firm normally functions and how it is to respond in unusual situations. While there is certainly a cost in preparing specifications, drawings, procedures, protocols, etc., the cost difference between doing an inadequate job and a superlative one is quite minimal and, in fact, usually has paybacks due to efficacy, reduced waste, and fewer failures.

Frequently Asked Questions

Q. *Our documentation group sees the benefit of using different formats for procedures, such as the multi-column format with diagrams, but our manager of QA feels they are too untraditional and won't meet the FDA's expectations. What should we do?*

A. There is nothing in any GMP that defines what a procedure should look like. Remember the purpose of the SOP is to be an informational tool to help people perform tasks safely, efficiently, effectively, and consistently. If a procedure is designed to accomplish that, your firm should be able to make that point to the inspector. Having a rationale included in a policy or SOP may help make that point.

Q. *What is structured writing?*

A. *Structured writing* is a generic term that encompasses such proprietary techniques as Information Mapping™. The basic concept is having a series of styles, or blocks, that are specific to particular uses, such as definitions, task descriptions, rules, etc. The content for each block is selected by a process called "chunking," which is identifying the amount of useful information that can be put into a block.

Drug and medical device firms that have used structured writing have found significant improvements in the efficiency of preparing documents as well as making them easier to use.

How to Write a Report

Mr. Tony Mattera, a corporate communications officer at Crestar Bank in Richmond, VA, has invented what he calls the "Jargon Master Matrix," a chart consisting of three columns of words that can be mixed and matched. Simply pick three numbers and use them to choose a word from each column to create phrases.

For instance, 11, 8, 7 give you a "value-based process model," while 1, 2, 9 yield "overarching support centralization"—all useful terminology for anyone trying to sound important (*The Wall Street Journal*).

1	overarching	visionary	objectives
2	strategic	support	alternatives
3	special	customer-oriented	expectations
4	specific	stretch	mechanisms
5	core	planning	assessment
6	long-term	marketing	update
7	defined	service	model
8	technology-based	process	product
9	formal	fundamental	centralization
10	exceptional	sales	incentive
11	value-based	budget	initiatives
12	executive	operating	feedback
13	immediate	discretionary	infrastructure
14	interactive	tracking	proposition

Chapter 14

Records and Reports

Goals

- Define *records and reports*.
- Identify the differences between records and reports.
- Discuss some issues in preparing records and reports.
- Identify items that should considered in creating and managing forms.
- Discuss the use of numbered forms and bound notebooks.

If functional documents, as discussed in the previous chapter, tell you what to do, records and reports are evidence to show what actually took place. They are the basis for reconstructing an event, decision, or task. They are created to fulfill business, scientific, and/or regulatory needs. One other function of records and reports is to demonstrate that the regulatory expectations were met while producing a drug or medical device.

Records and reports are produced according to documentation standards (discussed in the following chapter) in order that they can be a useful source of information without any question as to their integrity and authenticity. SOPs or methods usually are related to a record and describe how a specific record is to be completed. Records and reports are retained according to the applicable documentation retention procedures (Chapter 11).

What Are You Producing?

If you ask this question of most people in the pharmaceutical and medical device industries, they will immediately think of the tablets, pacemakers, surgical instruments, diagnostic kits, etc. that are shipped from their manufacturing site ultimately to the consumer. They are partially correct. The other item healthcare manufacturers make is *information*.

Information about the raw materials, manufacturing, testing, release, and a variety of other aspects describes how the product was produced. In GMP terms, if the GMP-required information is not correct or is incomplete in any way, the product, by definition, does not meet the GMPs. In other words, all the documentation must be correct and acceptable for the product to be released. *Batch release* (the drug term) or *acceptance* (the medical device term) is the process that includes document review, to ensure that all records are adequate.

Records

Records are documents that contain the raw data; data are entered directly onto a record. Records tend to be relatively simple documents with a limited focus. For example, cleaning logs, calibration work orders, lab worksheets, and sequential use logs are all records. A batch manufacturing or batch history record, although a more complex document, is also a record. Records answer the "who, what, when, where, why, how, and how much" questions that could be asked at some future time. Records can be paper- or computer-based.

Reports

Reports are more complex documents and provide more information than simply data—their purpose is to inform and, in many cases, be a call to action. They sometimes include a detailed discussion of what happened, including analysis of a problem or situation.

Reports might also be part of an information loop. That is, they may present recommendations based on a product complaint. These recommendations need to be acted upon in some way that will "close" the loop. This might mean taking action as recommended or documenting the rationale for *not* taking action. GMP auditors and agency inspectors frequently target reports (e.g., out-of-specification, deviation, complaint, failures) to see if the actions were completed in a timely way. Table 14.1 contrasts records and reports.

Table 14.1. Comparison of Records and Reports

Records	Reports
• Simpler	• More complex
• Data-oriented	• Information-oriented
• Descriptive	• Inform
• Have a historical, reference perspective	• May include analysis
	• May be part of a "loop"

What to Include

A dilemma in developing a template for a record or in writing a report is deciding how much information should be included. If you specify *too little* data to collect in a cleaning log, you may not meet the GMP requirements or, investigating a problem several months after the record was completed will be difficult. On the other hand, if you include *too much* data, it may make the record difficult to fill out. If spaces are left on a form, a deviation can result.

Two particular questions you can use to help determine the level of detail to include, particularly in records, are the following:

Why am I collecting these data? What is your purpose? Are you doing this to address a specific regulatory need, as defined by the GMP regulations or guidelines? If so, the regulations and guidelines may give you a minimum set of elements you will need to collect. Collecting data to answer a question that *might* be asked in the future is when people try to collect all the data they can. This is expensive and can cause a variety of problems. To prevent this, think through the most likely questions that will be asked. Is there some way of deriving information instead of actually collecting it? What is the risk in not being able to answer the questions? Talking to an experienced protocol or study designer or a statistician can help.

What will I do with this information? Once you get the information, what will you do with it? Might you be faced with a situation where you will have to take action that you would have preferred not to take? Is there a context in which the data points can be placed? For example, if someone decides to start examining rinse water to determine the level of a cleaning agent and discovers that it is 10 PPM, what does that really mean? Is it too high? Is it present in the final product? This does not mean you shouldn't collect information to improve the process or to better control it; it does mean that information can come back to haunt you. This needs to be considered at the outset and may mean collecting only what you absolutely require.

Records and Reports to Have

What records and reports should your firm have? The short answer is, *all the ones you need*. You determine this by reading through the various GMP regulations and guides and noting what is specifically required. One of the better listings is found in the European GMPs (EC, 1992) and in the WHO GMPs (WHO, 1992); they are similar and include specific information on what specific data elements should be included in various records.

In addition to the specifically required records and reports, you should determine what others will help you better understand and monitor activities and processes.

Forms

Forms are used to collect data. They can be simple, created using a word processing program and photocopied, or be complex, multiple page ones reproduced by a printer. In any case, forms need to be created and uniquely identified in a controlled manner. Table 14.2 lists items to consider when designing a form.

Table 14.2. Forms Design Checklist

Arrangement

1. Are all items, and groups of items, arranged in the right sequence, considering the source of the information, its use, and the way users normally write or read it?

2. Is the key information (that data used to retrieve the form from a file) in the most visible location?

3. Is there consistency between similar forms? (e.g., are signatures for review, approval, etc. in the same place?)

Spacing

4. Is the proper amount of space provided for each piece of information requested, considering the preparation method?

5. Is all extra space used proper for emphasis, separation, and balance rather than looking leftover?

Captions

6. Will all captions be easily understood by everyone who might use the form?

7. Are group captions or headings used to identify major areas?

Multiple choice answers

8. Are possible answers given whenever they would help the users provide better answers in less time?

9. Are the answers and check boxes arranged properly for the most productive and accurate fill-in?

Instructions and distribution

10. Can someone unfamiliar with the form complete it without referring to any other source for help?

11. Have all unnecessary instructions and explanations been left off the form, so that the user's intelligence will not be insulted?

(Continued on next page.)

(Continued from previous page.)

12. Are the distribution instructions shown on the form in the most effective way for its usage pattern and construction?

13. Do people know what to do with the form when it is completed?

Margins

14. Does the form have adequate margins for the required border on the photocopy machine?

15. Are the margins adequate for other handling characteristics, such as filing, copying, stapling, etc.?

Type, lines, and screens

16. Do all of the lines on the form do what they are supposed to do? (Some guide, some separate, some stop.)
17. Is screen (shading) used where helpful to separate, highlight, or identify fields or zones, not just to decorate the form? Will screens present problems if photocopying the form?

Identification

18. Is the title unique and meaningful?
19. Is the form properly and uniquely identified with a number for ease in referencing, ordering, inventorying, etc.?
20. Is the organization properly identified?
21. Is there a revision number or code?

General

22. Will this form accomplish its purpose with the minimum amount of effort by all users?
23. Is this the best possible tool to do the job?
24. Is the form visually appealing and inviting to the user?

Having forms approved *before* they are used may seem like a waste of time. There are several reasons why it is important. First, if the information collected on the form serves some GMP purpose, someone needs to consider if the appropriate amount of data will be collected, what will be done with the data, what will be done with the completed form, how long should it be kept, etc. Second, forms have a strange way of mutating—with the help of people. Someone might see a form and decide that it would be better if one extra field was added or if more space was provided for a response. If the form serves as a vehicle for getting data into a computer, there may be a problem if people are making a 50–character response instead of the 15–characters that the program was designed to accept.

Just like procedures and other functional documents, data collection forms need to have a unique identity. Part of this identity typically includes a version, edition, or release number. Versions of a form are sometimes linked to a procedure. For example, a heading on a form may be:

Sequential Log for Ribbon Blender RB-24

For use with Procedure OP435

Version 2 and higher

If a form is referenced in a procedure or method, a copy of it should be included as an attachment to the procedure.

Unnumbered vs. Pre-numbered Forms

Unnumbered forms are simply copies of an original—there is no difference between the first copy and the 874th copy. Pre-numbered forms, on the other hand, are like bank checks—each has its own unique number, typically in sequence. With numbered forms someone keeping track of the numbers (or accounting for the numbers after the fact) can determine if a particular form was used or turned in.

Quality auditors and regulatory agency inspectors do not like to see unnumbered forms in a GMP setting because of the potential for lack of control and nonaccountability. If someone writes a lab result on an unnumbered form and then decides it was a mistake, it is very easy to throw away the first form and fill out a second one. FDA inspectors, who have seen this frequently, take an interest; pre-numbered forms will allow an inspector to quickly see if there are proper controls on raw data collection.

Pre-numbering can be done using certain word-processing applications or by a commercial printer with a special form-numbering tool. Optionally, after the blank forms are printed numbering can be done using a mechanical stamp that automatically changes after each impression. If pre-numbering is done internally, there should be an SOP describing the task and also someone who is in charge of this. You may be asked to prove that no one else can recreate a pre-numbered sheet.

Treating forms as blank bank checks is a good way to think of the controls that must be placed on them. You should know to whom or to what process, test, etc. a form was assigned. If for some reason a form needs to be invalidated, this should be noted and signed-off by at least two people. A procedure should describe this process in detail.

When are pre-numbered forms most appropriate? Whenever data upon which decisions are to be made are recorded or when the form is the copy of record for testing or inspection data. Pre-numbered forms are not as critical when the information is simply for notification purposes.

Log/Notebooks vs. Forms

Logbooks and forms (either pre- or un-numbered) are both types of records, but there are situations when one is preferable over the other.

Logbooks are very useful when the data collected are limited to a particular item or activity, and you want to confirm that an activity has been completed or monitor it over time. For example, the sequential use log for a lab instrument would contain when the instrument was used, for what analysis, the person using it, and if anything unusual was observed. This makes it easy to track when problems might have started or, in the event of a calibration problem, what tests may need to be rerun.

Forms are useful when the data collected are more numerous or complex, or if a specific action is to be taken based on the data. For example, some firms use pre-numbered forms to document out-of-specification results. The investigation is documented, reviewed, approved, and signed-off upon by various lab and quality unit personnel. Forms are more useful when the information is self-contained, that is, the data do not have to be in context with other, similar data points.

There are times when logbooks and forms are used in a complementary way. For example, a logbook would be used to record what OOS investigation form was taken out, when, by whom, and when it was closed out.

Prebound vs. "Homemade"

One final decision about bound notebooks is whether they should be purchased premade with stitched bindings or whether they can be made internally using plastic or wire coil bindings. Your decision needs to be based on the potential use of the information.

If data in a lab notebook may be used for patent, legal, or regulatory submissions, prebound notebooks should be used that easily allow the data be shown to be complete and authentic and to withstand questions as to their integrity. In situations where the data or information are not as critical legally or from a regulatory perspective, having simple, internally-made notebooks will suffice. For example, these would be acceptable in most situations for cleaning logs or sequential use records for a filling machine.

One caution concerning loose-leaf binders: many auditors and inspectors are wary of these, as pages can be added or replaced easily. The more controlled your records are, the more confidence others will have in your system. This is especially important if auditors or inspectors are concerned with data authenticity and integrity.

Laboratory Notebooks

Lab notebooks have a special function as a collection point for data related to basic research and discoveries. Most R & D organizations have procedures that describe how the documents are to be used, reviewed, verified, archived, and maintained. Some general points to cover in your procedure are shown in Table 14.2. If lab notebooks may be used in making patent claims, your procedures and practices should be reviewed by a legal expert in this area.

Table 14.2. Items to Consider as Part of a Laboratory Notebook SOP

- Issuing and assigning of notebooks
- Accounting for notebooks issued and returned
- Documentation standards
- Review and signature standards
- Security of notebooks while in use
- Inclusion/filing of "attachments"
- Assigning keywords to notebooks for later search and retrieval
- Microfilming of notebook contents
- Archiving original notebook and microfilm

Failure Reports

Of all the internal GMP reports drug and device firms prepare, failure or deviation reports continue to be of significant interest to the regulatory agencies, particularly to the FDA. Auditors and inspectors expect that firms will document their problems, take the appropriate action on the lot(s) or product(s) affected, gather the "lessons learned," and use this information to prevent recurrences.

Failure reports can be brief or long, depending on the seriousness of the problem and the breadth of the investigation. In some cases, a succinct one-page document may be adequate; other cases, like a series of mysterious sterility failures, may take a hundred pages to document and discuss.

What should you have in a failure report? Some guidance comes from the FDA's Medical Device Quality System Regulation (FDA, 1996) as well as inspectional observations made by FDA personnel. Your firm's practice needs to be defined in a procedure. The failure report should include information about the following:

- Identity of existing and potential causes of the nonconforming products (i.e., failures);
- What caused the problem to be brought to light;
- Investigation to discover the cause of the nonconformities;
- Extended investigations made to determine if the problem has affected other products or batches;
- Actions needed to correct/address the particular nonconformity and prevent recurrences in the future;
- Success of the corrective action and that it had no other, unintended affects on the product or process; and
- Periodic review of the corrective actions to ensure that they were adequate.

The report needs to be approved by management and the quality unit; it must be distributed to management for their information and action.

Conclusion

Records and reports, including batch history records, use logs, lab notebooks, and worksheets, are the informational tools that are needed to document that activities were performed and decisions were made, it is hoped, according to the relevant procedures, specifications, methods, and GMP expectations. These documents need to be readily accessible for business and regulatory reasons.

Frequently Asked Questions

Q. *We are concerned about paper in our aseptic manufacturing areas and the potential of its causing microbial and particulate contamination. Is there anything special we can do?*

A. Firms are using a couple of methods to reduce this kind of contamination. The first is to use a special paper that is designed to shed fewer particles and can also be autoclaved. You print onto this paper (available either as sheets or in rolls from TexWipe and Kimberly Clark), and then autoclave the paper as it enters into the aspetic area. The second method that seems to be used more and more involves the laminating of the batch record. The laminated sheets are sanitized using an alcohol solution as they enter the aseptic area, and personnel record onto the sheets using felt-tip markers. Immediately after the production run, the sheets are photocopied and signed by all personnel involved, indicating that the photocopy is an authentic reproduction of the record that was generated. The laminated sheets are then wiped clean and reused. If this second method is used, it must be defined by an SOP.

Chapter 15

Recordkeeping Techniques

Goals

- Identify and define the nine characteristics of a properly prepared record.
- Describe the meaning of a person's signature placed on a GMP document.
- Discuss what should be included in a procedure on recordkeeping.
- Discuss training of personnel in recordkeeping techniques.

Agencies that regulate the drug and medical device industries all require a vast number of records and reports that are used in evaluating the product before it is approved to be marketed. Additional data are required to support the release of a lot into the marketplace. But, what are their requirements for what makes a proper, adequate record?

As with many other GMP issues, what agencies expect is not highly specified. The European Community and WHO GMPs give some idea as to their requirements:

.6 Where documents require the entry of data, these entries should be clear, legible, and indelible. Sufficient space should be provided for such entries.

.7 Any alteration made to a document should be signed and dated; the alteration should permit the reading of the original information. Where appropriate, the reason for the alteration should be recorded.

.8 Records should be made or completed when any action is taken and in such a way that all significant activities concerning the manufacture of pharmaceutical products are traceable.

(WHO, 1992)

Characteristics of a "Good" Record

What does a good or adequate record look like? In order to get people to prepare good records, they need to know specifics; saying that "you'll know it when you see it" is not quite specific enough.

A properly prepared record has nine characteristics which apply whether the document is being created for GLP (Good Laboratory Practices), GCP (Good Clinical Practice), or GMP reasons. Specific recordkeeping techniques are used to create a document that meets the nine characteristics; these techniques are discussed at the end of this chapter.

The nine characteristics are:

1. **Permanent**—the information cannot be changed, erased, or washed off.
2. **Legible**—the information can be easily read.
3. **Accurate**—calculations are correct and other information is recorded with care.
4. **Prompt**—information is recorded in a timely way. Actions are documented immediately after they are performed, never before.
5. **Clear**—everyone who reads the document has the same understanding of what it means.
6. **Consistent**—all information, such as dates, times, abbreviations, etc., is standardized so there is no confusion.
7. **Complete**—all required information is included.
8. **Direct**—information is recorded onto the proper form, notebook, or computer system without any intervening steps.
9. **Truthful**—all information included in the document is, to the writer's knowledge and ability, what really happened.

To see how these characteristics are incorporated into a record, let's take a closer look at each of them.

Permanent

In the drug and device industries, *permanent* does not mean forever. Rather, it means that the data and information will be available through the active and inactive life of the document. In some cases this can be for several years, while for critical documents it could be several decades.

For paper-based records, permanence is primarily achieved by using writing and printing methods that use indelible or water resistant inks. Pencils or erasable inks are not to be used. While watching an in-process package integrity test, an auditor was surprised to see technicians using water-based ink pens. A review of their work showed numbers that were smeared and only partially legible due to water droplets.

Another ink problem is with the photostability of some inks. At a workshop on Documentation Systems, one participant told of an experiment their firm did which showed that red ink (used exclusively by the QA group at the firm) faded much more rapidly than blue or black ink. Whether this would be a problem for documents needing to be available for seven years is doubtful.

Fax machines and some types of computer printers are notorious for creating nonpermanent documents. Older style fax machines and lab instruments using thermal paper can have their printouts deteriorate due to heat and light. Newer, ink-jet models usually have water-based ink that won't fade, but will run when wet. Photocopying these documents is a way to create a permanent record.

For laboratory notebooks using attachments, "permanent" means mounting the printouts, photographs, etc. in such a way that they will not be separated from where they are originally placed, either by accident or intentionally. Some adhesives and tapes are better than others for this. Stapling a document into a record is not generally done, as it can be rather easily removed and replaced.

Other aspects of achieving document permanence are discussed in Chapter 11.

Legible

Legibility is the characteristic most people think of when considering whether a record has been properly prepared or not. In a GLP/GCP/GMP context, it not only means that the printing and handwriting can be easily read, but that any error was properly handled.

Documenting according to GMP means never deleting anything, be it on a paper document or in a computer file. Changes, done according to a procedure, that include information on when the change was made, who did it, and why may be done, as long as the original entry can also be seen. A legible, properly corrected error does *not* mean using Liquid Paper, White-Out, or similar products, nor does it mean obliterating or erasing the original entry.

Accurate

Accuracy in recordkeeping includes double checking calculations as well as being sure all data and information are correct when they are being entered on the document or into the computer. One aspect of accuracy has to do with rounding practices and the use of significant figures; this is most applicable to laboratories and is usually defined in a lab SOP.

When it comes to lot/batch identity, accuracy is extremely critical as a transposition of a digit can lead to total confusion.

Usage of decimal points is also an accuracy issue. In several different hospital incidents, patients have been fatally overdosed because of a misplaced decimal point, such as the 0.9 that was given instead of the .09 that was intended.

Prompt

For a record to have legal and regulatory credibility, it must be made contemporaneously with an action or decision. You document in real-time, not in a batch mode. In discussions with operations and laboratory personnel, the meaning of *prompt* can become a debatable point. Part of this problem can be traced back to how methods and work instructions are "chunked" and broken into steps and substeps. If they are written with the workers in mind and how the tasks are actually performed, documenting the tasks becomes much easier.

Included in the definition of *prompt* is the prohibition against one's signing for something before it is done. One manufacturing firm learned that its people treated the sign-off points more as a field where they listed who on their workteam was to perform the assignment. When an unexpected fire drill cut short their workshift, workers on the following shift came in, and, thinking that the marked steps were complete, went on to the next set of steps. It was an in-process assay that indicated something was very wrong; the mistake cost the firm several hundred thousand dollars.

Clear

While *legibility* has to do with the readability of the words or characters, *clear* has to do with the meaning and understandability of what is written. Something can be very legible, but not meaningful.

In one sense, records are not created for their creators, but for others who will use the records at some other time and, possibly, in another place and with a very different point of view. It is essential that all readers of the record will come away with a common understanding of it. One test for the clarity of a document would be to think how easy it will be for someone five or ten years from now to understand what was intended.

Things that can improve the clarity of a record are use of standard words and abbreviations, making sure all units of measure are included, as well as the design of the document. People entering information should be sure to read their entries or have them read by a colleague or supervisor. It is much simpler to clarify a passage at the time it is written than to struggle with understanding what it *could* mean several years after the fact. A friend who kept a professional diary found this out when she was called into an investigation. Even though she kept detailed notes that were understandable when written, she had a difficult time being definitive as to what she meant three years after the notes were written.

Consistent

What is 01/03/04? Even if you know that it is a date, it could be at least one of three, depending on the format used. In the United States, it is likely to be January 3, 2004; for Canadians and Europeans (and those with a military background), it probably is the first of March, 2004. If a computer system gave the date, it could be the year 2001, March 4th. Unfortunately, "probably" is not a word inspectors or auditors like hearing. Similar confusion can exist with times: AM, PM, and 24-hour time systems.

As companies and information becomes more globally distributed, organizations and writers need to be specific and consistent to avoid confusion. Having a standard listing of abbreviations, units of measure (that includes how they are to be noted, e.g., k, K, Kg, or KG) is a place to start.

Complete

When you notice that data are missing from a record or report, you wonder if the data were collected in the first place, or if the writer simply forgot to enter it. An informal study at a large drug manufacturing firm showed that resolving incomplete batch records was one of the biggest reasons for delaying release of products.

What should be done when something is missing? The person who should have entered the data is the best person to address the problem. If that person is not available, some firms allow a supervisor or quality unit representative to sign-off on the entry. An auditor's question would be, *How do you know that the step was completed? What do you have to prove that it was done as described?* An explanation would be expected as to why the entry was not made at the time and the rationale as to why it can be appropriately made at a later time.

Direct

This characteristic of a properly produced document relates to where the data are first entered. All data must be placed into the official record or file. By default, the place where data are entered becomes the official record. FDA inspectors continue to make observations about personnel (those in the QC/analytical labs most frequently) writing information on the back of envelopes, Post-It® notes, and even their hands. One drug manufacturing firm, when being audited by an outside auditor, tells of one formulator who wrote a weight on the pants leg of his white uniform. The auditor, wanting to make the point that the pant leg was now the official record, insisted that the section of fabric with the number be cut out and attached to the batch record.

Truthful

The records and reports associated with the development of a product and the manufacture, testing, and release of a batch or item need to describe *what really happened*, not what should have occurred or what was expected. There have been some unfortunate cases in the early 1990s when company officials were less than truthful about the information they provided to regulatory agency personnel. At one firm, this went as far as keeping two sets of batch manufacturing records—one that showed the product was made according to the approved application, the other documenting how the product was *really* made.

In reality, while an extremely small percentage of individuals and firms would attempt to be dishonest in what they document, auditors and inspectors, particularly those from the FDA, take a suspicious view of all of the industry. Should they get even a *hint* that something is amiss, their inspection will take a very different tack as they try to determine if the records are authentic and truthful.

All personnel who create records and reports *must* understand that they have a professional, legal, and ethical responsibility to prepare documents that have integrity. They need to describe what really happened and when, even if it may show weaknesses in the firm's quality system.

One of the "little" places where proper judgment is needed is in dating of entries. If an entry was missed when the record was first produced and caught during a review, what date should be used? The more tempting approach, because it is fast, is for the reviewer to just put in what should have been the original date and sign for the person who should have signed the record. Even with the best of intentions, for a rather trivial entry, this is an improper act—falsification of data would be the criminal charge that could be made against the person doing it. *All* personnel need to know what they should do and that their management will tolerate nothing less.

Signatures and Initials

What does it mean when a person signs or initials a GMP document? What legal and regulatory implications does it have? While the words *signature*, *signatures*, *sign*, or *signed* show up in 733 FDA documents or regulations, nowhere are those words defined.

An article written by the FDA's internal expert on electronic identification, Paul Motise (1992), presented the FDA's viewpoint on this. According to the FDA, signatures have three primary functions. Signatures

- Identify the actor (person doing the work, making the decision, etc.) and show his or her responsibility to act;
- Document the action in a way that is legally binding and cannot be altered or repudiated; and
- Create a record that is admissible evidence in court.

Initials are also used in documents. From a practical point of view, they are the same as signatures.

A problem in many documents (and not just those used in GMP environments) is that a signature line is provided for the person to sign, but no specifics are given as to what that signature means. For instance, an insurance agent asked a person to sign a form with the agent orally telling the person that it just meant that an inspection took place. However, nowhere on the form was that meaning conveyed. Someone might conclude the signature meant that the person agreed with the conclusions or was willing to settle.

One of the best elements in the Electronic Signature Regulation 21 CFR 11 (FDA 1997) is the requirement that whenever a signature is required, records must contain information about the meaning that is associated with that signature. This can be a simple line such as "Investigation conducted by" or "Work done and reviewed by." This approach should be used in all GMP documents, be they paper-based or electronic.

If electronic signatures are allowed, the FDA requires that each person certifies they understand the meaning of the electronic representation of their signature. (See Chapter 2 for more detail.)

Your Procedure for Recordkeeping Techniques

Before you can teach people how to prepare proper records at your facility, you need to define what you expect. For example, how do you want dates recorded? Must they be in U.S. formats or European formats? Can they have the numerical form of the month (1, 2, 3) or the first three characters (Jan, Feb, Mar)? Most all firms want their people to document things correctly, but many have not specified what *correct* means.

A recordkeeping procedure identifying your requirements is an important GMP SOP. While not asked for in the regulations, it becomes part of the firm's standard way of doing business. Things that should be included in the SOP (along with an example) include:

- Type of writing instrument (one-piece retractable ball-point without a separate cap or pocket clip);
- Color of ink (either black or blue);
- Format for recording of dates (either Mon Day/yr [May 26/98] or Day Mon yr [26 May 98]);
- Format for recording time (24 hour clock);

- Acceptable, standard abbreviations (mg, L);
- How to handle blank spaces ("NA", initial, date);
- Initialing and signing records (what your signature, initials mean);
- Attaching/including nonstandard documents into a batch record or laboratory notebook (sign across the edge and then tape using non-yellowing, permanent tape);
- Error correction—immediate (one line through the original entry, write corrected entry above or close to original entry, provide a reason if not obvious, then initial, date); and
- Error correction—discovered later (original person must be found to make correction; if not, supervisor is permitted to do so; justification for making the correction at the time; signature, date when correction was made).

Training Personnel in Recordkeeping Skills

One of the most important training courses personnel working in a GLP or GMP facility should take is on recordkeeping techniques. It is not a difficult skill to learn, provided the expectations are defined in a procedure. Everyone needs to understand why documentation is so important and that it needs to be done all the time.

One packaged course, *Documentation Skills and Techniques*, is available in both English and Spanish; firms can adapt the materials to accommodate their own recordkeeping standards. (See the Resources section for information on obtaining the program.)

Conclusion

One of the best ways of showing an auditor or agency inspector that your firm is serious about GMP is to produce records that are of high quality, meaning that they meet the nine characteristics discussed in this chapter. This cannot be done unless people understand what is required of them and are trained and constantly encouraged to properly document their work.

Frequently Asked Questions

Q. *What color of ink should we use as we document? Our manufacturing people prefer blue, but we've heard the FDA requires black.*

A. There is nothing in the GMPs that states what color you must use; certain inspectors have their own preferences. The historical reason for black ink is that it was easier to photocopy (sometimes blue inks resulted in light images); also black ink tends to be more stable and resistant to solvents. As photocopier quality has improved, some FDA personnel have preferred blue ink as it can indicate whether a document is an original or a black and white copy. (Color copies are another dilemma . . . .)

Whatever your firm decides to be acceptable, be it black, blue, or both, it should be written down in the procedure; this makes your practice easier to defend if an inspector challenges you.

Q. *Why should we define an "official" pen?*

A. Some firms have learned the hard way that certain pens can cause problems. For instance, pen caps have a way of coming off and ending up in a mixer, resulting in rejection of the batch. One-piece stick pens with retractable points that are sealed are a good solution.

Hospital Errors Related to Poor Documentation and Recordkeeping Skills

Not writing things down.

Duplicate medication was given to a patient by a senior floor nurse and an orientee. Both believed that they were responsible for giving medications to the patient. Neither the nurse nor the orientee wrote down they had given the patient the medication (Davis and Cohen, 1981).

Illegible handwriting.

An order was sloppily written before angiography for Seconal 100 mg IM. This order looked like Demerol 100 mg IM. Demerol was administered. The procedure was canceled one day and subsequent surgery was postponed until the next week. Hundreds, if not thousands of dollars [in medication costs] were wasted by this delay, not counting the increased anxiety which must have been suffered by the patient (Davis and Cohen, 1981).

An unclear decimal point.

A patient suffering an acute attack of gout was seen by a physician. The physician ordered "colchicine 1.0 mg IV now." The decimal point was not seen by the nurse reading the order and she interpreted the dose as 10 mg. A syringe was prepared from stock and administered to the patient. The patient eventually died from an overdose (Davis and Cohen, 1981). (Note: For labels, the USP prohibits drug manufacturers from using a zero following a decimal point, e.g., 1.0 mg is not permitted.)

Chapter 16

Moving From Paper to Electronic Documents

Goals

- Discuss the use of electronic document systems in the drug and medical device industries.
- Define some commonly used acronyms and terms related to electronic systems.
- Identify and discuss what to consider before making the move to an electronic system.
- Suggest ways of staying current with industry trends.

It is an understatement to say that information technology has changed our work and environment and the ways we communicate. While some dominant trends are emerging, no one can really predict where things will be ten years from now. Even our language does not move fast enough: it took years for the term *automobile* to replace *horseless carriage* or for *radio* to take the place of *wireless*. Who knows what the new term will be for electronic batch records. Marshall McLuhan said it best: We are "looking at the future as though through a rear-view mirror."

While some drug and medical device companies have adopted and integrated electronic systems into their development, regulatory, manufacturing, laboratory, and quality operations, most firms in the late 1990s are only in the early stages. This chapter does not give many answers. Rather, it identifies issues to consider and suggests ways to make the migration from a paper to an electronic environment a little easier.

Because technology is changing so quickly, the information included here is of a general nature; suggestions are given at the end of this chapter of periodicals and sources for keeping up with the electronic publishing industry.

The Alphabet Soup of Electronic Systems

To begin the discussion of electronic systems, we need to be talking about the same things. There are a variety of terms and acronyms used; some are more prevalent in drugs while others tend to have more of a niche in the device arena as shown in Table 16.1.

Table 16.1. Terms and Acronyms Used Regarding Electronic Systems.

DMS	Document Management System—an application that supports the generation, review, approval, routing, and control of documents.
EDMS	Electronic Document Management Systems—see DMS.
HTML	Hyper-Text Markup Language—commands used in preparing text and graphics for use on the World Wide Web.
LIMS	Laboratory Information Management System—typically used for managing lab data, but many systems also include provisions for managing, distributing methods, and specifications.
MES	Manufacturing Execution System—an application that provides detailed instructions and data-input forms; forms the core of an electronic batch record system.
PDF	Portable Document Format—a proprietary method for creating documents that has an Acrobat® reader. Documents using this standard have a *.pdf* suffix.
PDM	Product Document Management—see DMS.
SGML	Standard Generalized Markup Language—a complex, but powerful international standard for creating the content and look of complex documents.
WSIWYG	What You See Is What You Get—the ability to see on a monitor exactly what is to be printed in the final version, be it on paper or on the Web.

What an Electronic System Is *Not*

If you are using a personal computer to prepare a procedure that is used as a paper document, is that considered an electronic documentation system? No. The FDA and most of industry considers that the equivalent of a smart typewriter and filing system (FDA, 1997, Comment #22, page 13437). Only when you have a system that manages the document, including its creation, modification, storage, distribution, and access are you in the realm of an electronic system.

Options

There is a wide variety of options available for document management systems. Your specific needs, types of documents (e.g., text or drawings), whether the system will be local or company-wide, the number of users and what they can do, and budget will determine the systems that are best for you.

The functions of collaborative generation, review, and approval tend to be what you pay the most for. The more capabilities you have and the more highly customizable these features are, the more you will pay for a system. Also, the more capabilities a system gives you for creating documents as objects, made of definable, dynamic parts, the more costly the system. The advantages you get from this need to be carefully weighed with the initial and ongoing costs.

There are some systems that are "off-the-shelf" solutions that have a fair number of capabilities and can be modified to your situation. At the other end of the spectrum, there are systems that are highly customized to a firm and require a significant installation process.

One of the more interesting and expanding areas for documentation systems uses internet/intranet technologies. Some large firms use a document management system to create, control, and manage the documents and then allow users to access them from the company's intranet system. Tools that create web-viewable documents are being integrated into basic word-processors, such as Microsoft Word and Correl's WordPerfect.

Before Making the Move . . .

The money spent on a documentation system's software and hardware is a relatively small portion of the overall cost of implementing an electronic solution, particularly if there are legacy documents (i.e., documents that were created in the past on older systems, or that just exist in paper-form) that you want to incorporate into the new system. Also, training costs are a significant piece of the equation. Therefore, you need to consider a number of things before saying "yes" to any system.

Understand your goals

What are you trying to accomplish with the new system? Is it to save time, costs, or headcount? Or, are you trying to provide documents that are more uniform looking and standardized in terms of content. Or, are you simply trying to stay at the leading edge of technology? Expectations should be made explicitly; priorities should be given to your goals, including the needs that are "must-haves."

Create a requirements document

Before you start using your system, you will need to assure yourself, your management, and the regulatory agencies that your system *really does* what you think it is doing: You will need to validate it. The specifications for your system need to be formalized in a requirements document. The requirements should be written in functional and understandable language; it should not be in "geek speak." The requirements documents don't need to be long; 5–20 pages should be the goal, as much more than this is difficult for any one person to understand.

Simplify the existing system

A favorite quote of those implementing new electronic technologies is, "Computers help you do stupid things faster." Computers also can cause chaos to occur much more quickly than people-based systems can. Before you look to put documents on an electronic system, you first need to understand your present, "as-is" system. You can easily do this by preparing a flowchart for each type of document that is used in your firm. You will need to note who does what, in what sequence, and how different pieces of information travel to different people. This takes time, but if you can't define your current system and what works and does not work, you will never be able to achieve the return on investment you are expecting with the electronic system.

The more complex parts in any documentation system are the document review and approval routings. Another complex part are the permissions given for reading, modifying, and creating documents.

Don't import your problems

One firm, as it was moving into an electronic environment, had its people simply recreate their existing batch records into the new system, using the same formats, use of redundant signatures, and non-value-adding verifications. If you are committed to doing things differently by using an electronic system, why recreate all of your same problems? It will take time and thought to improve your documents, but if you do it carefully and involve those who are the experts at doing the tasks (operators, technicians, etc.), the investment will have a large payback.

Don't think of forever

Even the best electronic solution with a well-established vendor will last only a decade—assuming the vendor keeps the application supported. More likely, changes due to hardware, software, and communications will make the solution's useful life closer to 5–7 years. Don't expect a solution to last forever.

As you look at solutions, be careful not to lock yourself into a highly proprietary system. If the firm goes out of business or decides not to support the product, you will be trapped on the proverbial desert island. Look for systems that have a significant marketshare across industries. Look for applications that have filters or conversion routines that can convert your documents into other formats.

Think "paper-lite," not "paperless"

People in the drug and device industries used to talk about "paperless manufacturing." This is a great concept, but, in the real world, you will be fortunate if you can simply reduce the amount of paper generated, resulting in a "paper-lite" solution. A conversation with several drug company executives illustrated the point:

they ask their secretaries to print out all their e-mail messages every day so they can read them more easily; the time they set aside for reading large SOPs and regulatory documents is when they travel by plane and don't have other distractions.

The more easily accessible electronic documents are, and the more value people can get by using only electronic documents (e.g., through hypertext links), the better your chances of reducing the amount of paper used.

Make the media work for you

In the early days of television, before writers and directors understood the power of the transmitted image and sound, programming consisted of placing a television camera in among the audience and turning it on. Later, directors discovered the power of different angles and positions of shots, the close-up shot, the use of hard cuts and dissolves. Over time, creators and viewers became more and more literate in the language of television.

The same is true with documents delivered using the computer. Reproducing an 8 1/2 by 11 inch document for use on the screen might be easy for the writer, but not for the user who has to continually scroll down and across. (An interesting parallel is seen with printing: While Gutenburg received credit for making the printing press that used movable type, printed documents didn't really become popular until Aldus started printing documents in a size that was more manageable—they fit into the side saddlebag on a horse.)

As you set up styles and formats for use, consult with people who understand interface design and those who can help you get the most from your documents. A relatively small investment here will make your documents much more accessible to those using them.

Staying Current

As mentioned earlier, things related to electronic documentation systems are changing rapidly. Some ways of keeping current are listed below. Consult the Resources section for phone numbers, addresses, and URLs.

- **Journals and trade magazines.** Don't limit yourself just to drug or device titles. *Publish* is a magazine aimed at graphics professionals, but includes a great deal of information related to publishing documents on the web and across platforms.
- **Trade organizations.** One trade organization, AIIM, the Association for Information and Image Management, has a number of publications and "state of the industry" reports.
- **Internet.** Almost all firms involved in electronic documentation systems have a presence on the World Wide Web. Commercial search engines can help locate where they can be found. Also, some industry journals like *BioPharm* and *MMDI* print listings of web-based resources.
- **Informal industry networks.** Since more and more drug and medical device companies have electronic documentation system initiatives planned or in place, finding one person who is involved may connect you with many others. Talk to your colleagues at other firms or at conferences.
- **Vendors.** Vendors of systems can help you stay current through their salespeople and literature, and at conferences they sponsor. Careful listening to a variety of vendors can make you an in-

formed consumer and educate you to ask the right questions.

Conclusion

Electronic document systems can help a firm create and manage documents more effectively and collect data to be used in improving processes. No system will fit your organization perfectly. In some cases, you will need to decide what will change the most: your own documentation system or the software itself. The best solution is one that optimizes both.

Books Have Endured for a Reason . . .

Ergonomics. This is probably the biggest difference between print and computers. It takes much manual dexterity to operate a mouse, buttons, scroll bars and the like, Books, by comparison, are simple to use.

Weight and Portability. Books weigh as little as a few ounces, but a good laptop computer with decent battery storage weighs 8 or more pounds. Yes, one could argue that a computer can hold hundreds, if not thousands, of books simultaneously. But if I am reading a novel at the beach this weekend, I don't care about the 599 unused novels in my computer.

Durability. Hit your friend over the head with a book. Now ask him to hit you over the head with his laptop. The book (or your friend) is more than likely to emerge from the collision unharmed than is the laptop (or you).

Intangibles. Read a book, close it, leave it on a shelf. No matter how much time may pass, the contents sill await you, unchanged. Electronic data, particularly on the Internet, are ephemeral and prone to unexpected editing or complete disappearance . . .

Power Consumption. Need we say more?

New York *Times*, May 25, 1997

Police in Manistee, Michigan are searching for the man who sold them the Crime Tracker software that was designed to help keep track of case records. The program, described by police as user friendly, simple, and comprehensive, has been eating years' worth of data.

Toronto *Globe and Mail*
Nov. 11, 1996

Chapter 17

Change Control and Documentation

Goals

- Define *change control.*
- Differentiate between changing a process and changing documentation.
- Describe a process for change control.
- Identify special requirements for medical device manufacturers.

Of all the elements present in a quality system, change control (also known as configuration management) is one of the most complex. The difficulties are not just in establishing a useful system, but in making it work without its collapsing on its own weight.

Change control is the formal process for proposing, evaluating, documenting, and implementing changes in the materials, components, equipment, facilities, methods, or controls involved in any aspect of producing, testing, labeling, or controlling a regulated product.

Change control attempts to minimize any surprises that might occur and help assure that the product meets all regulatory, customer, and quality requirements. Without change control, processes seem to take on a life of their own and follow the third law of thermodynamics: left on their own, things enter a state of greater and greater randomness.

Change control also considers the effect of the change on the validation status: minor changes may not require any additional validation whereas significant changes might.

Beginning in the mid-1990s, the FDA and the pharmaceutical industry have created and issued guidance to the industry related to changes. The series of documents, *Scale-Up and Postapproval Changes*, referred to as SUPAC, covers an expanding list of product types, such as solids and nonsterile semisolids. SUPAC guides are helpful in defining each change and in justifying and implementing it.

Change Control and Documents

How does change control relate to your documentation system? How does change control apply to revising a procedure or a specification? Before we can answer these questions, we need to focus on *what is really changing*.

There are two sets of things that can change in a GMP process: first, something about the activity itself changes—what is done, the equipment used, its sequence, who does it, where or when it happens. This is a *content change*; these changes would be proposed, evaluated, etc. by the formal change control process. SUPAC guidance documents discuss content changes.

Second, the change may be a *document change*. This occurs when the way the information is presented changes. A document change could be due to a content change, for instance, if the sequence of steps changes, or it could be done to make the document more understandable, for example, if one step was revised, breaking it into two steps. These changes must be controlled, but not to the degree that content change must be controlled.

If a document change alters the way the task is done or decisions are made, by definition that change is no longer simply a document change, but a content change as well.

Is there a ever a content change made without a corresponding document change? There could be, for instance, if the level of detail of the content change was finer (i.e., more specific) than that shown in the procedure or method. For example, if the method for turning on an instrument changed from pushing a button to turning a switch, and the SOP said simply "turn on the instrument," it would be a content change without requiring a document change.

Document Changes

Why distinguish between these two different categories of change? The distinction presented here is done to simplify and speed up the process of change control. It is based on the premise that certain changes to documents do not require that technical experts be involved in reviewing and approving them because the changes do not alter the decisions that are made or the manner in which activities are performed. At the same time, however, you need to ensure that important changes and corrections are not implemented without proper review. The method described here provides some balance of the two.

An important note: If your firm is covered by the FDA's Quality System Regulation or by ISO-9000, the standards require that the functions who initially approved a document be involved in the review of it, "*unless specifically designated otherwise*" [emphasis added] (ASQC, 1994). If your firm uses a special document change control practice as described here, your quality manual and related procedures must be very specific in describing the document changes that are permitted.

Some examples of document changes are the following:

- Format change,
- Style change,
- Simple rewrite of a paragraph, and
- Noncritical typo correction.

If the format or style change involves a modification to your SOP on document writing, for instance, requiring a change to your SOP on preparing specifications, that would be a content change that could, depending on your change control practices, require formal approval from your change control committee before it is implemented.

If the change is a simple correction, say, putting a previously prepared specification into the new format, the particular specification would not need to go through the formal change process. The same is true if someone discovered a misspelling or another noncritical typographical error.

A *noncritical* typo is a simple misspelling or a missing word. A *critical typo* is an error in an important aspect of the content. For example, if a specification called for the mixing time for a liquid to be "7–140 minutes" instead of the actual "7–14 minutes," that would be a critical typo that should be reviewed by a content expert before the changed document is approved. These types of change need to be defined in your documentation procedure.

Content driven document changes

As mentioned above, there are two types of document change: one initiated based on a content change, the other begun for other editorial reasons. Both types of change need to be described in a procedure.

When a document is changed because of content changes, the process used to revise the document needs to involve content experts and others, such as regulatory personnel, validation experts, etc. Some firms define the reviewers for changes as part of the change evaluation process. (Note: As mentioned above, the U.S. regulations that cover medical devices require that functions who initially approved the document be involved in reviewing changes.) The quality control unit should be involved in the review and approval process.

When an editorial change is made, there also should be a defined review and approval pathway. The quality control unit should have a role in the approval of these changes as well.

Documenting Changes

An expectation of any change control program is that changes are proposed, justified, and documented. This needs to happen if the changes are simple editorial ones or more significant content ones. A form is useful in communicating and recording the proposal, justification, and the decision that is made. These records would be kept as part of the document's historical file. Another place changes are often recorded is on the document itself.

For instance, a procedure may have a chronological list of the changes on the last page, or perhaps a mention of the change on the first page, with a chronological listing kept in the files. Another method is to asterisk ("*") the section of the document that was changed. Figure 17.1a-b provides examples of two approaches. Whatever is done needs to be defined in your procedure.

From a user's standpoint, it is useful to know very quickly what has changed. Having the change mentioned at the beginning and/or highlighted in some way is helpful.

Figure 17.1a. Chronological listing of changes.

Rev. No.	Effective Date	Description
4	May 17/97	Typo correction, section 4.3.d.
3	May 12/97	Addition of verification substep, section 4.3.d.
2	Jun 13/96	Change in temperature requirements for washing solution, section 4.1.b.
1	Feb 7/96	Initial version.

Figure 17.1b. Change mentioned at the head of a procedure.

Areas involved	This procedure applies to al: Apex Drug manufacturing sites worldwide that manufacture catalexin sulfate.
Reason for this revision	To change the temperature of the cleaning solution to be consistent with that which was the validated. (See Change Control Request 97.43.)
Implementation	This procedure shall be implemented at . . .

Revision Numbers

Each time a document is reissued, it needs to be distinguishable from the previous version. Release (or version or edition) numbers (or letters) can do this. This is discussed in detail in Chapter 5.

Saving Changes

In order to reduce the number of revisions, some documentation professionals will save up the changes and revise a document when either a content change is required or when there are a number of document changes that warrant the work in revising and re-distributing a document. This works well when the "trigger" for preparing a new document is described in the SOP, and there is a methodical way pending changes are held for action.

Conclusion

Change control is a critical element of your firm's quality system. Having a well-defined set of steps that people can use to propose and implement different categories of change will make the system work much more efficiently.

The Attempt at an Error-Free Book

The University Press in Glasgow, Scotland, attempted to produce the first perfect book. The work was the Foulises' editions of classical works, well regarded by nineteenth-century scholars and collectors.

Six experienced proofreaders were employed to check each page meticulously. After that the pages were put up in a university hall for two weeks, with the notification that 50 pounds sterling would be given to anyone who could find an error on any of the pages. Only then, when the publishers believed every possible error had been discovered and corrected, was the work issued.

But upon publication it was discovered that the book contained several errors—including one in the first line of the first page (Goldberg, 1984).

CHAPTER 18

Making Your Documentation System an Investment

Goals

- Describe what an audit is.
- Provide the business reasons/benefits for conducting an audit.
- Identify things that could be examined during an audit of a documentation system.
- Discuss costs and benefits related to a documentation system.
- Identify ways of reducing costs and increasing benefits of a documentation system.

With your GMP documentation program in place and working, you're finished, right? There's nothing else you need to do as long as SOPs, specifications, methods, and the like tumble out of the printers. There *are* a couple of additional elements, though, that are expected of a well run system. First, you and your management will want to be sure the program meets its regulatory and procedural requirements. Second, you will want to find ways to improve the program that save money and make documents more useful to those using them.

Auditing the Documentation System

An audit is a

> documented, systematic tool, used by management and done periodically by independent, qualified people to verify and evaluate an organization's use of the principles of good manufacturing practice and compliance applicable GMP regulations (Vesper, 1997).

An audit of a documentation system could include examining the entire system and how it operates or it could focus on the functional document and record/report side. If done properly, an audit should show how well you are meeting the regulations and your internal requirements. If problems are observed, someone will need to identify corrections that will resolve the particular issues. Figure 18.1 shows a checklist specific to documentation systems.

There are many purposes for doing an audit, one of which includes ensuring that the system meets all regulatory requirements. This needs to be done on a periodic basis and on special occasions, for instance, far in advance of an agency pre-approval inspection.

Figure 18.1. An audit approach for documentation systems.

Primary question to answer

Is there an integrated process of developing, approving, controlling, distributing, and archiving functional documents (e.g., specifications, SOPs, protocols, lab methods) and completed records (e.g., logs, batch history records)?

(Continued on next page.)

(*Continued from previous page.*)

General things to look for

- Formalization of the system (e.g., SOP).
- People have the correct documents when they need them.
- Documents are controlled with ID and revision numbers.
- Documents have been reviewed and approved before use.
- Users do not have to make choices about which documents should be used.

Specific examples—functional documents

- SOPs, methods, etc. are available where and when they are needed (i.e., at the "point of use").
- Documents have been reviewed and approved and training has been provided *before* they are effective.
- People are performing the tasks according to the SOPs, protocols, and methods.
- Uncontrolled copies of documents are not used.
- Master folders/files of documents are arranged systematically, and archived information is readily available.
- Documents are written clearly with the users in mind.
- Documents are reviewed periodically for their accuracy, relevance, and compliance to GMP expectations.
- Record retention period, archiving facilities.

Specific examples—records and reports

- Records and reports meet the nine characteristics:
  - ⇒ Permanent
  - ⇒ Legible
  - ⇒ Accurate
  - ⇒ Prompt
  - ⇒ Clear
  - ⇒ Complete
  - ⇒ Consistent
  - ⇒ Direct
  - ⇒ Truthful
- Records and reports can be accessed quickly.

Opportunities for Improving Your System

As the definition of audit indicates, the audited item is compared against specific standards, e.g., regulatory requirements. Your system may be meeting these standards, but it may have other deficiencies that are costing time, money, or difficulties for those using documents. In other words, there is room for improvement.

If you consider your documentation system as an investment, there are two basic ways of improving the return on that investment: reducing the costs and increasing the benefits. Usually, firms work on both sides of the equation at the same time.

Documentation Costs

The cost of your documentation program is a combination of visible and "invisible" or hidden costs. The visible costs include the salaries of the people involved, equipment, supplies, and workspace. These are the costs that management usually tries to reduce, by trying to increase productivity, adding new electronic tools, etc. Why? Because they are easily visible and are the traditional ways of squeezing dollars.

The invisible costs, those that people do not see or do not *want* to see, as they can be even more staggering than the visible costs, include the following:

- Time of subject matter experts to prepare documents;
- Time/resources for document review;
- Time/resources for document approval;
- Re-re-re-revising documents;
- Storage costs (file cabinets, floor space) for retaining "in-process" documents;

- Time/resources spent tracking down documents that were lost in the review/approval cycle;
- Time spent reformatting documents into standard formats;
- Having multiple formats for documents;
- Distribution of unneeded documents;
- Wasted materials, time spent redoing work due to use of an out-of-date document; and
- Wasted materials, time spent redoing work due to an incorrect or hard-to-understand document.

From this incomplete list, you can see that many of the points are time related—time wasted in doing what is nonvalue adding: waiting, tracking down, looking up information, redoing, and revising. Some of the highest value items can be wasted materials and products.

Documentation Benefits

Some of the benefits of a documentation system are harder to quantify using actual numbers than others; some benefits are very "soft." When it comes to wasted materials, cost savings can be easily measured. The reactions of users to documents is another "soft" item to examine. It can be reasoned, however, that the easier, more accessible a document is, the more likely it is to be used. Following are some examples of benefits, listed in order from easiest-to-measure to less easy.

- Increased productivity (due to information, documents that are easier to use);
- Decreased storage costs (file cabinets, floor space) for retaining active documents;
- Decreased storage costs (file cabinets, floor space) for retaining inactive documents;

- Less time spent looking for active documents, records, reports;
- Fewer regulatory observations that have to be replied to;
- Better, faster decision-making;
- More consistent task performance and decision-making;
- Fewer deviations that need to be investigated;
- Decreased time spent in training; and
- Having multiple uses for information that is included in documents.

Adding Metrics to Your Documentation System

Before you can decrease your costs or increase the benefits, you need to be able to measure them. It is important that you have an idea of the current situation, so you can determine the significance of your cost savings and quality improvement activities.

For example, what statistics do you have on your documentation system? Do you know:

- How many original documents are generated each year?
- How many revised documents are processed each year?
- The number of revision cycles each document requires (not just averages)?
- The length of time it takes the average document to go through the system?
- The optimal time it takes for a special ("emergency need") document to go through the system?
- How many people use a given document?

- The average number of changes made during a revision?
- Who are the people/departments with the most frequent revisions per document?
- Who takes the longest to review a document?
- The amount of time documents just sit in "queue" waiting to be processed further?
- How long it takes to retrieve a given document?

These are just some of the statistics that are useful in monitoring your system. If you can connect these with a financial indicator (e.g., dollars), you will have information that management will notice. You will have taken some of the hidden costs and made them visible.

For example, one large firm in North Carolina reported at a technical conference that, on an annual basis, the time it takes their personnel to find and access internal information (SOPs, batch records, development reports, etc.) is worth $950,000. That's a lot of money to waste each year! This figure (based on a methodical investigation) got the attention of their management, who invested a significant amount in networked information systems. Another firm discovered that one of the biggest costs was in writing functional documents. They significantly reduced these costs, and improved document quality, by training all of their personnel in structured writing techniques.

Failures and other horror stories

If you are trying to discover the problems and costs of inadequate documentation, ask those who use it—they will have horror stories that can be used to get management's attention. For example, if people have had recent incidences of failures, rejections, reworks, etc. that were due to unclear procedures, improving the instructional documents would be a cost avoidance. Many times the failure investigation reports will blame the problem on inadequate employee training. Experts who have read hundreds of these are quick to say that many, if not most, of these problems have poor documents as a contributing cause.

Lost opportunities

As you think of the time people are spending rewriting, re-reviewing, and so forth, what could they be doing instead that is more profitable? Maybe your organization is having to hire temporary staff or outsource certain activities because your internal personnel are too busy with documentation. Here, time and potential opportunities are being lost.

In this case, you want to show the benefits of having a simple system, one that will encourage doing things properly the first time. Also, tools and training may help reduce the amount of time spent, making people more productive. For example, think of the time saved if your can reduce document churning? Or the time spent by subject matter experts as they search for the correct format to use in writing a specification or preparing a drawing.

Putting Together a Plan for Improvement

What can you as a documentation professional do to improve your current system? There are some relatively simple steps you can take for identifying a problem and correcting it. As with other GMP systems, your change control process will need to evaluate the ramifications of any change before it is implemented.

1. **Know the current situation.** Document what is going on now. Flowchart the big picture, flowchart how specific documents are created, reviewed, approved, and distributed,
2. **Discover where the bottlenecks and problems are.** Statistics on time and multiple handlings of documents will show this. Also, problem solving tools like fishbone diagrams, problem maps, and other graphical representations are extremely helpful.
3. **Look for the golden opportunities.** The Pareto Principle says that 80 percent of the problems are caused by a very few factors. Look first at these. Where will you get the most bang for your effort?
4. **Identify what *success* is.** If you make an improvement, what will it look like? How will you know when the improvement has been achieved? Categorize what you want to accomplish in terms of "must-haves" and "nice-to-haves." This will help in setting priorities.
5. **Get people involved.** Let people who are involved and feel some of the frustration and pain be part of the solution; they need to use whatever improvements you come up with. Form a team to help with the project.
6. **Develop the proposed solution.** Map out what you want to do. Make drawings on long strips of wrapping paper. Ask people for suggestions. If they say something won't work, don't let them go until they give you a way to improve it. For complex problems, you may want the help of an outside expert/consultant.

7. **Refine the proposed solution.** Take the feedback you receive, and as a team, come up with a better way of accomplishing your goals. Make sure you are getting all of the "must-haves" that were identified.
8. **Build in evaluation tools.** This will be important to show that the project is working. The more quantitative you can be, the better. However, don't discard people's reactions.
9. **Have the solution approved.** This step would involve review according to your firm's change control process.
10. **Do a pilot project.** Limiting the scope of your initial work will generally be more acceptable to management who want to reduce the impact of any failure. (If you've done a good job in developing the solution, the risks of failure will be low.) More positively, the pilot project will provide information that you will learn from and be able to apply to the enterprise-wide solution.
11. **Evaluate the pilot results.** What went well? How can things be improved even more? What will you do differently next time?
12. **Make plans for implementing a larger-scale version.** Here, you want to standardize your solutions to other areas.

Conclusion

Almost every documentation system is brimming with opportunities for improving it, with reduced costs and better, more useful documents as the results. Don't be tempted to just try technology as the solution: you need to understand the underlying root causes. Maybe these can be solved using technology, or maybe there are better, simpler ways of addressing the issues. As you work for improvement, remember that your documentation system is an investment and that the paybacks are not just financial, but include meeting the regulatory and quality goals of the business.

The Most Embarrassing Bible Printing Error: "Thou Shalt Commit Adultery"

In 1631, a printing of the Bible was found to have no negative in the listing of the seventh of the Ten Commandments. It flatly told the reader, "Thou shalt commit adultery."

As a result of this error, the Bible came to be known as the Wicked Bible. Its printers, Robert Barker and Martin Lucas, were fined 3,000 pounds (Goldberg, 1984).

GLOSSARY

Note

The terms and definitions selected are those that may be of interest to people involved in documentation-related activities. Sources for the information include White, 1988; Adobe, 1997; and Rainwater Press.

A more complete glossary of printing and publishing terms is available at the Rainwater Press website: *http://www.rainwater.com*.

Acid-free paper	Paper that contains no acid or acid-producing chemicals that cause it to yellow.
Active record	A document that is used or is being frequently accessed.
Anamorphic scan	Scanning an original piece of artwork so that the width and height are not enlarged or diminished in proportion, but are altered to produce an image that is taller and skinnier or shorter and fatter than the original.
Auxiliary storage	Storage of data in a device outside the computer's own internal storage.
Back matter	Pages following the main text of a publication, devoted to reference matter such as appendices, glossary, and index.

Backup	Copy of work or information saved in case of later loss or damage to the original.
Baseline	Imaginary horizontal line on which the bottom of the main part of the letters is arrayed. The descenders of the letters "g,j,p,q,y" drop beneath it.
Basis weight	The weight in pounds of a ream (500 sheets) of paper cut to a standard size according to the grade of paper.
Batch processing	Processing where similar tasks are delayed and grouped for combined handling.
Bit-mapped type	Dots representing the shapes of characters that are stored in the digital typesetter's memory and are called up as needed by a beam of light.
Bleed	An element, usually an illustration or screened area, printed to the edge of the paper.
Boilerplate	Standard stored text, elements of which can be rearranged and combined with fresh information to produce new documents.
Bond paper	A grade of paper that is stronger and more durable than the average sheet of paper, often used for letterhead.

Book paper	A grade of coated or uncoated paper used in books.
CAD/CAM	(Computer-Assisted Design/Computer-Assisted Manufacturing) System used to produce graphic output.
Caliper	The thickness of paper, usually expressed in mils (thousandths of an inch).
Callout	Wording placed outside the figure but attached by an arrow to the part of the illustration to which it refers.
Central file	A central location where documents are deposited and available for use/review.
Character	A character refers to a single letter, punctuation mark, number, space, or any other object or symbol in a typeface set.
Compatibility	A characteristic of equipment permitting one machine to use the same information or programs as another without conversion or modification.
Copy	Wording of a manuscript. Also, in printing, all the material to be printed.
Copy of record	The "official" version of the document that is subject to the document retention policy. (This presumes that other copies that have been generated have been properly

	destroyed after use.) The copy of record may not be the "original," e.g., the firm's copy of record of a letter sent to the FDA may be a copy retained by the person (or department) who wrote it.
Custodian	The person (or title) in the organization responsible for holding a particular document.
Decentralized file	A location, close to where documents are generated or used, that contains the document.
Disposal date	The point in time when the document is no longer needed and can be destroyed. The disposal date would be included in the document retention schedule.
Document center	The location where inactive documents are stored and maintained.
Document inventory	A methodical audit, resulting in a detailed report, of what documents are kept by whom and in what form of media.
Document retention schedule	A detailed plan for where and how long particular documents are to be kept before being destroyed.
Document series	A group of a particular type of document (e.g., SOPs, specifications) that are stored together.

Dot	Unit of measurement representing 0.08 mm of 1/300 inch; also known as a *pixel* or *spot.*
Dummy	A preliminary layout showing the size, shape, form, and general style of a printed piece.
Em	The square of the type size: in a 10-point type size, the em measures 10 points high and 10 points wide.
Em-dash	A long dash (—), used primarily to indicate a break in thought.
Emoticons	Small graphical renderings of emotion composed in ASCII characters. Sometimes called *smileys*. Includes a grin :-) shock (:-o) and a stomach in knots <:/& Emoticons are generally not suitable for use in communication with regulatory agencies.
En	Half of the em: in a 10-point type size, the en measures 10 points high and 5 points wide.
En-dash	Short dash (–), slightly longer than a hyphen (-), used instead of the words *to* or *through* (e.g., letters "A–Z").

Expert set	An expert set, or expert collection, is a font that has a more refined, or expanded, set of typographic characters than most other fonts. Expert sets may contain old style figures, picture fonts, embellishments, fractions, or other unique characters.
Fixed pitch	Typeface in which all character cells are of equal width.
Folio	Page number.
FSL	(Forms Source Language) Programming language used to design forms.
Grain	The direction in which the fibers in a sheet of paper have been made on a paper machine.
Halftone	Printable version of a continuous-tone original such as a photograph. Grays are converted to dots of varying sizes to simulate dark or light tones.
Inactive record	A document that is not being used but must be retained for regulatory, business, or other archival purposes.
Kerning	The process of adjusting the space between letters to produce visually consistent text. Some combinations of characters naturally have excessive space between them

	(such as Ta or Vo) and must be manually adjusted by the designer or typographer.
Leading	Refers to the amount of space between lines of text. This term's origin is from the bygone era when little strips of lead were inserted between the lines of type. Leading and kerning can both have an impact on the readability of text (both positive and potentially negative, if incorrectly applied).
Microfiche	Sheet of film carrying miniaturized images for storage. Usually 105 mm x 148 mm (4.1 x 5.8 inches) accommodating 98 8.5 x 11-inch pages at 24x reduction in 7 rows x 14 columns.
Orientation	Printing on a page in the portrait (vertical) or landscape (horizontal) mode.
Pica	A unit of measurement commonly used in typography. One pica = 1/6 of an inch.
Pixel	PICture ELement) smallest addressable unit of a bit-mapped screen; also known as a *dot* or a *spot.*
Point	Another unit of measurement commonly used in typography. One point approximately

	equals 1/72 of an inch. There are 12 points in a pica.
PostScript font	A font designed to conform to the PostScript page description language. These fonts require a PostScript printer to render accurately or a non-PostScript printer with Adobe Type Manager installed. PostScript fonts are actually composed of 2 subunits called screen fonts (or bit-mapped fonts) and printer fonts (or outline fonts). The PostScript category includes both Type 1 and Type 3 fonts as well as multiple master fonts. PostScript fonts are scaleable and are preferred by most service bureaus and used for most high-end output.
Preview	Preliminary view of a document on a display terminal before printing; also known as *soft proof.*
Proportional scan	Enlargement or reduction of an original whose height and width remain in proportion to each other (*compare* Anamorphic scan).
Purge	Eliminating those documents that no longer have business, regulatory, or archival value to the organization.
Ream	Five hundred sheets of paper, a standard measurement.

Recto	Right-hand page.
Resolution	Fidelity of reproduction, depending on number of spots-per-inch. High resolution (with more spots) is finer and more precise than low resolution, which yields coarser and rougher-edged results.
Screen font	One of the two components of a PostScript font. These are created by sending electronic "bundles" of information to certain pixels (or dots) on the computer screen thus allowing you to view the font on-screen. Also known as bit-mapped font.
Scrolling	To cause an element displayed on the screen to move up or down vertically, or side to side horizontally, to show more than can fit in the space on the screen or a window on the screen.
Text font	Another appearance-based or usage-based category of fonts. Text fonts are used for body copy and are most commonly serif fonts. In large families of typefaces these are often denoted with the variants "regular" or "book" (e.g., Times Regular, or Kabel Book).
Text paper	Any fine-quality printing paper.

TrueType font	A font format/technology jointly developed by Apple and Microsoft that is in direct competition with the PostScript font format. Like PostScript fonts, TrueType fonts are scaleable. Unlike PostScript fonts, TrueType fonts have only a single component and do not require a PostScript printer or Adobe Type Manager.
Typeface family	A collection of all sizes and styles of typeface that share a common name and common characteristics, e.g., Times Regular, Times Bold, Times Bold Italic, etc. are all members of the Times Family.
Variable text	Keyboarded or prerecorded text (such as addresses) added to standard recorded text to complete a document.
Verso	Left-hand page (the reverse side).
Vital documents	Those documents that are essential to an organization's survival. Would typically include certain financial records (e.g., contracts, accounts payable), research notebooks, and product-specific documents.
Weight (1)	The thickness of a line or typeface.

Weight (2)	The density of paper measured in pounds.
Weight (3)	The relative darkness of the characters in the various typefaces within a type family. Weight is indicated by relative terms such as thin, light, bold, extra-bold and black.
Word processing	Entry, editing, storage, and accessing of words using a keyboard or terminal and magnetic storage.

—References and Readings

Adobe. 1997. *TypeReference*. San Jose, CA: Adobe Inc.

ANSI/ISO/ASQC. 1994. *Guidelines for Auditing Quality Systems*. ANSI/ISO/ASQC 10011-1994. Milwaukee, WI: ASQC Press.

Cohen, Michael. 1989. *Nursing 89*. October 1989, p. 18

Cohen, Michael. 1991A. *Nursing 91*. November 1991, p. 22.

Cohen, Michael. 1991B. *Nursing 91*. May 1991, p. 22.

Crystal, David. 1987. *The Cambridge Encyclopedia of Language*. Cambridege, UK: Cambridge University Press.

Davis, Neil, and Michael Cohen. 1981. *Medication Errors: Causes and Prevention*. Philadelphia: George F. Stickley Company.

Diamond, Suan Z. 1991. *Records Management,* Second Edition. New York: Amacom.

EC. 1992. *The Rules Governing Medical Products in the European Community*, Vol. IV, Good Manufacturing Practice for Medicinal Products. Luxembourg: Office for Official Publications of the European Community.

Eulenberg, Julia N. 1986. *Handbook for the Recovery of Water Damaged Business Records*. Prairie Village, KS: ARMA International.

FDA. 1996. Title 21 Part 820—Quality System Regulation for Medical Devices. *Federal Register*, Vol. 61 No. 195, October 7, 1996.

Fields, Timothy J. 1995. "Inventorying SOPs/Policies Against Regulatory Requirements" in *PharmTech Proceedings*. Eugene, OR: Advanstar Publications. pp. 5–12.

Flower, L., Hayes, J. R., and Swarts, H. 1983. "The Scenario Principle." In P.V. Anderson, R.J. Bockmann, and C.R. Miller (Eds.), *New Essays in Technical and Scientific Communications: Research, Theory, and Practice* (pp. 41-58). Farmingdale, NY: Baywood.

Gold, Gloria. 1995. *How to Set Up and Implemnt a Records Management System.* New York: Amacom.

Goldberg, M. Hirsh. 1984. *The Blunder Book*. New York: Quill/William Morrow.

Hollis, Thomas, and Barbara Meserve. 1994. "Electronic Documentation Routing Methods" in *PharmTech Proceedings*. Eugene, OR: Advanstar Publications. pp. 465-494.

HPB. 1996. *Good Manufacturing Practice Guidelines*, 4th ed. Ottawa, Ontario, Canada: Health Protection Branch Drugs Directorate.

Kodak. 1996. *Storage and Preservation of Microfilms, Document D-31*.

Kodak. 1997. Kodak Website. *http://www.kodak.com* Frequently Asked Questions-0700.

McRobb, Max. 1989. *Specification Writing and Management*. New York: Marcel Dekker, Inc.

Motise, Paul J. *PharmaTechnology* November 1992, pp. 29-35, 77.

NFPA. 1990. *Fire Protection for Archives and Records Centers (ANSI/NFPA 232AM)*. Quincy, MA: National Fire Protection Association.

Nygren, Wm. V. 1980. *Business Forms Management.* New York: Amacom.

Panati, Charles, 1984. *Browser's Book of Beginnings.* Boston: Houghton Mifflin Company.

Redish, J. C. 1989. "Reading to Learn to Do." *IEEE Transactions on Professional Communication*, 32, December 1989, pp. 289-293.

Skupsky, Donald S. HD, CRM. 1991. *Recordkeeping Requirements.* Denver: Information Requirements Clearinghouse.

Vesper, James L. 1997. *Auditing: Clear and Simple.* Buffalo Grove, IL: Interpharm Press.

Vesper, James L. 1993. *Training for the Healthcare Manufacturing Industries*. Buffalo Grove, IL: Interpharm Press.

Wheildon, Colin. 1995. *Type & Layout.* Berkeley, CA: Strathmoor Press.

White, Jan V. 1988. Graphic Design for the Electronic Age. A Xeron Press Book. New York: Watson Guptill Publications.

WHO. 1992. WHO Expert Committee on Specifications for Pharmaceutical Preparations. Thirty-second Report (includes WHO GMPs). Geneva: World Health Organization.

Wurman, Richard Saul. 1990. *Information Anxiety*. New York: Bantam Books.

Appendix 1

Document Examples

Note

The following documents are based on examples I have seen used in the pharmaceutical and medical device industry. They are shown to point out good and poor practice. They are not meant to be "perfect" documents, nor are they meant illustrate correct content.

Example 1A. Specification—Initial Version

This example is modeled after a specification used by a drug manufacturer. It is missing a number of expected items such as quality unit approval signature or statement and date. The bottom date is ambiguous . . . is it when the document was issued? Approved? Effective?

The test specifications ("It must meet the USP test") are links to another document that the analyst must get.

The form itself is poorly identified; there is neither a version nor a release number on the document.

(ID ####S)

MANUFACTURED RAW MATERIAL SPECIFICATIONS

(Not for use in customer or purchasing negotiations or in product registrations.)

Title: WATER FOR BULK STERILE OPERATIONS AND FINAL RINSE

Description: It is a clear, colorless, odorless liquid. It is water purified by an approved method. It contains no added substance. It is USP quality Water For Injection except for the Particulate Matter requirement. It meets the USP requirements for Microbial Limits.

pH: Between 5.0 and 7.0 by the USP method.

Total solids: Not more than 10 ppm (0.001%) by the USP method.

Oxidizable Substances: It must meet the USP test.

Chloride: It must meet the USP test.

Sulfate: It must meet the USP test.

Ammonia: Not more than 0.3 ppm by the USP method.

Calcium: It must meet the USP test.

Carbon Dioxide: It must meet the USP test.

Heavy Metals: It must meet the USP test.

Microbial Limits: Not more than 10 organisms per 100 mL

Bacterial endotoxins: Not more than 0.25 USP Endotoxin Unit per mL.

November 8, 19__

1.00

Example 1B. Specification—Revised Version

This is a revised version of the specification shown in Example 1A. It is laid out to be easier to read and takes advantage of bold typefaces.

Signature lines have been added, as well as a page number (1 of 1), effective dates, and a caution statement that links this document to a material safety data sheet (MSDS).

The test method and criteria for acceptance are more clearly laid out than in the previous version.

Acme Fine Pharmaceuticals	Identification code	**####$**
Material Specifications	Name	**Water for Bulk Sterile Operations**
	Effective date	May 5, 1993
	Revision number	3

Grade: Does not apply. Empirical formula: H_2O

Appearance: **Clear, colorless, liquid.** Formula weight: **18.0**

Other descriptors: **Odorless.** Sample storage conditions: **5-10 deg. C.**

Expiration/re-evaluation date: Does not apply.

Note: ####$ is USP Water For Injection except for the Particulate Matter requirement.

Caution: **Hot liquid. May cause burns to unprotected skin. See MSDS ####$**

	Tests (and method) to be used	Criteria for acceptance
1)	pH (USP method)	Between 5.0 and 7.0.
2)	Total solids (USP method)	Not more than 10 ppm (0.001%).
3)	Oxidizable substances (USP method)	________ (meets USP test).
4)	Chloride (USP method)	________ (meets USP test).
5)	Sulfate (USP method)	________ (meets USP test).
6)	Ammonia (USP method)	Not more than 0.3 ppm.
7)	Calcium (USP method)	________ (meets USP test).
8)	Carbon dioxide (USP method)	________ (meets USP test).
9)	Heavy metals (USP method)	________ (meets USP test).
10)	Microbial (LM 183)	Not more than 10 organisms per 100 mL.
11)	Bacterial endotoxin (LM 652)	Not more than 0.25 USP Endotoxin Units per mL.
12)	Particulate matter (LM 635)	____________________.

Approved by (and date): ______________________________

Approved by QC (and date): ______________________________

page 1 of 1

Form 35.2 NOT FOR USE IN CUSTOMER OR PURCHASING NEGOTIATIONS OR FOR PRODUCT REGISTRATION.

Example 2. Batch Record Cover Sheet

The batch record cover sheet conveys to manufacturing personnel that the batch record can be used to make the product. It also summarizes the approval for release made by the quality unit when they do their final review.

The document gives some guidance as to what signatures mean, but it could be improved by, for instance, a more specific meaning of what *approved* really means.

Criteria for certain items (e.g., red QC approval stamp) is useful, as is the reference to the process validation and the most recent change control order.

MBR $$$####$$ Lot Number ####$$#

Vial Number ID123, Acmencin, 50 mg per 2 mL, 2 mL
(Excelmycin Sulfate Injection USP)

Theoretical Yield —126,532 vials

Form Number 2

Pages: 25	Manufacturing Department: PR23
Revision Number 3	Mfg Ticket issued by: ___________
Revision Date: 09/23/91	Yield: ___________
Master formula approved by TS: MLB	Coupons inspected by: ___________
Input by: JLT	Approval for inspection by: ___________
Document editor: MT-V	Preliminary approval by: ___________
	date: ___________
Validation reference: 90-75	Final QC approval by: ___________
	date: ___________
LLS closed by: ___________	Final QC approval stamp (must be in red)
date: ___________	

Reason(s) for most recent revision: (1) Minor format changes, as per SOP 4963. (2) To revise upper accountability limit from 101. % to 100.5 % (Change Control Order 91.37).

This manufacturing ticket has been reviewed by technical services and has been approved for use only if signed and stamped below.

Approved by:

Date of approval:

Technical services stamp (must appear in blue):

Document printed 1:44 PM, 7/8/93. page 1 of 25

Example 3A. Batch Manufacturing Sheet and Work Instructions—Initial Version

This document was used by a manufacturing firm as the instructions and data sheet for producing a cleaning solution. It is similar to the other batch records that operational personnel filled out.

There are many problems here. The use of capitalization makes it very difficult to read, as do the breaking up of the steps using a column of opening parentheses.

There are no spaces in which to record the actual amounts added.

The line on the left is assumed to be where operators initial, but there is nothing specifying this.

All in all, this is a very difficult record to use and to review during the batch release document review.

MBR ###$$$ Lot Number ###$$$

Part I

Solution Manufacture
(Clean Area)

Group leader: ________

________ Date process started: ________ Time: ________.

Sulfuric acid reagent (####$) checked into area by: _______

________ Prepare sulfuric acid solution 25% (W/V) as follows:

________ (NOTE: SULFURIC ACID SOLUTION 25% W/V MAY BE PREPARED NO MORE THAN 7 DAYS BEFORE REST OF TICKET.)

SOLUTION A

________ (ADD AT 20 - 25 DEGREES C TO A CLEAN PROPERLY LABELED
(20 LITER PYREX BOTTLE PLACED IN A STOCK POT:
(
________ (WATER FOR INJECTION (ID ####$) ************** 8 KG 567 G
(
________ (OUTLET NO. ______ DATE: ______ TIME: ______
(
(ADDITION VERIFIED BY: ______
(
________ (ADD VERY SLOWLY AND CAUTIOUSLY ACCORDING TO
(SOP 746:
(
________ (SULFURIC ACID REAGENT (ID####$) ************************
(************************************ (1 L 32 ML) 2 KG 84 MG
(
________ (ADDITION VERIFIED BY ______
(
________ (TO AID COOLING TRANSFER SOLUTION TO CHILLROOM AND
(MIX THOROUGHLY TO COOL SOLUTION TO 25 DEGREES C OR
(BELOW.
(TIME STARTED: ______ DATE: ______
(
________ (TIME COMPLETED: ______ DATE: ______
(
(VERIFIED BY: ______

SOLUTION B

________ (ADD AT 20 - 25 DEGREES C TO A CLEAN PROPERLY LABELED
(20 LITER PYREX BOTTLE PLACED IN A STOCK POT:
(
________ (WATER FOR INJECTION (ID ####$) ************* 5 KG 436 G

Document printed 1:44 PM, 7/8/93. page 2 of 25

Example 3B. Batch Manufacturing Sheet and Work Instructions—revised version

This example is a revision of the preceding batch record. There are some improvements, both in the content provided as well as in the layout.

More whitespace is used as well as rules (boxed borders) to set off certain sections.

The document has two ways of giving specifications—ranges (e.g., 20–25 degrees C) and "calculatables" (e.g., ±0.03 kg). Ranges are preferred for ease of use and consistency.

There are no spaces to record the actual amounts added.

The use of arrows and italics makes the form a little busy, but could help the users.

The lines on the left are not identified as to their purpose; it is assumed that operators write their initials there.

MBR ###$$$ Lot Number ###$$$

Part I — Solution Manufacture
(Clean Area)

Group leader: ______________

Date process started: ________ Time: ________

Sulfuric acid reagent (####$) checked into area by: _______

Preparing Solution A [sulfuric acid solution 25% (W/V)]

Equipment used	The items listed below will be used. All equipment and utensils must be clean and identified. (Cleaning SOPs: 486 and 488; Equipment identification SOP: 243.) • 20 L Pyrex glass bottle (placed in a 10 L stock pot for added stability). • Tygan hose for WFI transfer. • Protective equipment for acid handling.
Procedures used	The following procedures are called for (in addition to the routine operational procedures used). • Handling Acid and Caustics (SOP 523)

Processing steps

_______ 1.1) Set WFI heat exchanger to 20-25 degrees C. Wait until WFI is at 20-25 degrees C.

_______ 1.2) Add Water for Injection (ID####$) to 20 L Pyrex glass bottle

→ Amount to add: **8.56 Kg** (±0.03 Kg)

WFI outlet No: ______ Date: ______ Time: ______

Addition verified by: ______

_______ 1.3) Add Sulfuric Acid Reagent (ID####$)

Consult SOP 523; wear protective equipment.

Add acid slowly and cautiously.

→ Amount to add: **2.84 Kg** (±0.02 Kg)

Addition verified by: ______

_______ 1.4) Transfer solution (now called SOLUTION A) to chillroom and mix thoroughly to cool.

→ Required temperature: 25 degrees C or below.

Started Date: ______ Time: ______

Finished Date: ______ Time: ______

Temperature/finish time verified by: ______

Document printed 1:33 PM, 7/8/93. page 1 of 25

Example 4. Sequential Log

This is an example of a cleaning log located in a manufacturing room. The items recorded are based on the GMP requirements.

The first line is a "cue" or example line, showing what format the data is to be placed in. This would help people remember what is required.

The record also includes a "reviewed by" check, presumably done by a supervisor after all entries have been completed.

The original form does not have any control number (e.g, a form number) nor is there anything that links the form to a specific SOP.

Acme Fine Pharmaceuticals
Tank Cleaning Log

Please complete as shown on the first line.

Date	Tank ID	Time – **start**	Time – **finish**	Past product	Past lot number	Process used	Cleaned by	Inspected by	Anything unusual ? (if yes, note and use form DR1)
Jan 4/97	WW-35	10:35	13:15	Lidorex	XC352	SOP42.2	[illegible]	[illegible]	NA - [illegible]

Completed form reviewed by: ______________________ Date: ____________

Example 5. Procedure Review Coversheet and Comment Sheet

This coversheet would accompany the draft SOP as it is being reviewed. It is a summary of who reviewed it and also if any changes were required.

If changes were required or there are any comments, they would be recorded on the comment sheet.

Neither form has a control and/or release version on it. It would be helpful to note what SOP defines its usage.

The signature space does not include what signing means.

Acme Fine Pharmaceuticals	Procedure Review Coversheet
Procedure title: ____________	Procedure no. ____________
	Proposed rev. no. ____________
	Review cycle no. ____________
Revision Code: ____________	Change control no. ____________
Proposed effective date: ____________	**DUE DATE:** ____________

Reason for revision:

Reviewers:

Title	Name	Signature	Date	NO changes req'd	**Changes required***
Manager, Tech Services	Kurt Wilson				
Manager, Production Ops	Linda Murray				
Manager, QC	Steph Allen				

*** All changes and comments must be written on the attached sheet.**

Acme Fine Pharmaceuticals	Procedure Review ***COMMENT*** Sheet page ___ of ___
	Procedure no. ____________
Procedure title: ____________	Proposed rev. no. ____________
	Review cycle no. ____________
Revision Code: ____________	Change control no. ____________
Proposed effective date: ____________	**DUE DATE:** ____________

Page	Section**	Suggested change	Rationale	Initials

** Please also highlight the words or sections with a yellow highlighter and return the document to QA Documentation Services.

Example 6. Template for a Bilingual Procedure

This template was used by a manufacturer who wanted to have their critical procedures in both the official "corporate" language and in the local language.

Note how the approval signatures are built into the header of the front page. On the second page (and each succeeding page), the approvers need to re-initial. If they miss one, it would be considered a deviation and the procedure could not be used. This is an example of how good intentions for control can get carried away, causing problems.

<table>
<tr><td rowspan="6">Acme
Fine
Pharmaceuticals

PROCEDURE
PROCEDIMIENTO</td><td colspan="2">SOP No./Rev. Code: PA-009A
Numb Procedimento:</td></tr>
<tr><td>Location:
Planta:</td><td>(Plant name)</td></tr>
<tr><td>Department:
Departmente:</td><td>Packaging Ops.</td></tr>
<tr><td>Effective:
Effectidad:</td><td>3/93</td></tr>
<tr><td>Supersedes: 1.6
Supersedo:</td><td>Date: 12/91
Fecha:</td></tr>
<tr><td>Preparer:
Preparado:
______
QA SOP</td><td>Date:
Fecha:
______</td></tr>
<tr><td rowspan="3">Title: Procedure issuance

Titulo: (Title – 2nd language)</td><td>Reviewer:
Rev:
______
Dept. Manager</td><td>Date:
Fecha:
______</td></tr>
<tr><td>Approvals
Apprebado
______
Safety Manager
______
QA Manager</td><td>Date
Fecha
______

______</td></tr>
<tr><td>Page of
Num de</td><td>Pages
Pagina</td></tr>
</table>

PROCEDURE - Language #1

I PURPOSE

qweroiuyqweriouyertqw;kjhef;guqwig; owhsrghil;lingrl;liqryuio;e

II RESPONSIBILITY

qweroiuyqweriouyertqw;kjhef;guqwig; owhsrghil;lingrl;liqryuio;e

Procedure - Lanuage #2

I PROPOSITO

zxvnm,,.bv;lkjgjkl;dsopoitursa/kfgjkl sdfhfoeirhidirgihrwiioywihlf

II RESPONSABILIDAD

zxvnm,,.bv;lkjgjkl;dsopoitursa/kfgjkl sdfhfoeirhidirgihrwiioywihlf

Doc ID Number, revision number, file name

Acme Fine Pharmaceuticals	SOP Title SOP Titulo:		**Procedure issuance**
SOP No./Rev Code Num de.../Codico.... **PA-009A**	Effective Date Feche de Effectidad **3/93**	Supersedes Date Superdeda Fecha **12/91**	Page Number Nom. de Pagina **2 of 9**
Originator/date Orginator/Fecha	Department Mgr./date Gov. de Departmento	Safety Manager/date Goveme de Securidad	QC Manager/date Goveme de QA

PROCEDURE - Language #1, continued

III Definitions

qweroiuyqweriouyertqw;kjhef;guqwig; owhsrghil;lingrl;liqryuio;e

IV Procedure

1.0
qweroiuyqweri ouyertqw;kjhef;gu qwig;owhsrghil;lingrl;liqryuio;eqweroi uyqweri ouyertqw;kjhef;gu qwig;owhsrghil;lingrl;liqryuio;eqweroi uyqweri ouyertqw;kjhef;gu qwig;owhsrghil;lingrl;liqryuio;e

2.0
zxvnm,,.bv;lkjgjkl;dsopoiturs a/kfgjklsdfhfoeirhidirgihrwiio

Procedure - Lanuage #2, continued

III Definiciones

zxvnm,,.bv;lkjgjkl;dsopoitursa/kfgjkl sdfhfoeirhidirgihrwiioywihlf

IV Procedimiento

1.0
zxvnm,,.bv;lkjgjkl;dsopoiturs asdflkjf;lkretudfgjiklsdafjfdidirgihrwiio zm,mg,a,mdfbkghklddskjgjkl;dsopoitur s a/kfgjklsdfhfoeirhidirgihrwiio w;kjhef;gu qwig;owhsrghil;lin

2.0
srghil;lingrl;liqryuio;e srghil;lingrl;liqryuio;eqweroiuyqweri ouyertqw;kjheqweroiuyqweri ouyertqw;kjhergihrwiio

Doc ID Number, revision number, file name

Appendix 2

Resources

Those involved with documentation systems may be interested in the following publications, organizations, and materials. This list is not intended to be vast and inclusive, but rather, things the author has found useful.

Document Retention Time Requirements

- *Recordkeeping Requirements*
- *Records Retention Procedures*
- *Legal Requirements for Microfilm, Computer, and Optical Disk Records*

Publications by or edited by Donald S. Skupsky, JD, CRM and available through:

Information Requirements Clearinghouse,
3801 East Florida Avenue, Suite 400
Denver, CO 80210, USA

(303) 691-3600

- *Guide to Record Retention Requirements*

U.S. Government Printing Office,

Washington, DC 20402

Style Guide

WIRED Style–Principles of English Usage in the Digital Age. Edited by Constance Hale. 1996, San Francisco: HardWired.

GMP Training Materials

Documentation Skills and Techniques

(courseware with video, instructor's guide)

LearningPlus, 1996.

Phone: +1 (716) 442-0170

Fax: +1 (716) 442-0177.

Internet: http://www.learningplus.com

Internet-based Resources

Note: These site addresses (URLs) were valid at the time of publication. Using a web-based search engine (e.g., Yahoo or AltaVista) is a method of finding other current sites.

Canadian HPB Website:
http://www.hwc.ca/hpb

U.S. FDA Website: ***http://www.fda.gov***

Regulatory Affairs Website:
http://www.medmarket.com/tenants/rainfo/master.htm

Rainwater Press (Glossary of printing terms):
http://www.rainwater.com

Emergency Response and Salvage Wheel

National Institute for Conservation
3299 K Street, NW, Suite 602
Washington, DC 20007-4415

Phone: +1 (888) 979-2233
Fax: +1 (202) 625-1485

Industry and Professional Associations

Electronic Documentation Systems

Association for Information and Image Management International (AIIM)

1100 Wayne Avenue, Suite 1100

Silver Spring, MD 20910 USA

(301) 587-8202

Records Management

Institute of Certified Records Managers

P.O. Box 8188

Prairie Village, KS 66208

Magazines

Publish–The Magazine for Electronic Publishing Professionals.

Publish

P.O. Box 5039

Brentwood, TN 37024

1-800 656-7495 or

615-377 3322

About the Author

James L. Vesper

James L. Vesper designs and develops instructional courses and workshops for the pharmaceutical and medical device industries. He established and is president of the firm, LearningPlus, Inc. and has had more than 18 years experience in the pharmaceutical industry.

Mr. Vesper worked eleven years at Eli Lilly and Company, Indianapolis, Indiana. His first assignment was as corporate industrial hygienist, followed by three years in Corporate Quality Assurance. There, he was responsible for issues concerning the manufacture and testing of parenteral products made at Lilly facilities and third-parties worldwide. His last assignment was as Project Leader of GMP (Good Manufacturing Practice) Education and Instruction, establishing the department and its mission.

Since 1991, Mr. Vesper has been creating innovative instructional products for the pharmaceutical and health-care industries using selected video and computer technologies as more effective and efficient delivery media. Working as consultants with a wide variety of clients, his firm creates integrated curricula for personnel and customized training courses targeted to particular needs. He speaks and writes for various international technical and professional organizations, including ISPE, GMP TEA, PDA, Pharmaceutical Sciences Group, and PharmTech.

In addition to traveling extensively and working on-site in Singapore, Latin America, India, France, England, Puerto Rico, and Italy, Mr. Vesper has been on the faculty of Indiana University, teaching courses in Industrial Hygiene and guest lecturing at other universities. He has appeared on stage in the Metropolitan Opera's production of *Aida* at New York's Lincoln Center in a supernumerary role.

Mr. Vesper earned a B.S. in Biology from Wheaton (Illinois) College and a Masters of Public Health from the University of Michigan School of Public Health in Ann Arbor. He is a member of ASTD and PDA.

James L. Vesper
LearningPlus, Inc.
1140 Highland Avenue
Rochester, NY 14620 USA

Phone: (716) 442.0170
Fax: (716) 442.0177
Email: *jvesper@learningplus.com*
Website: *http://www.learningplus.com*

INDEX

D

J

K

L

M

N

O